Heavy Metal Contamination and Its Effects on Ecosystems and Human Health: Challenges and Solutions

Heavy Metal Contamination and Its Effects on Ecosystems and Human Health: Challenges and Solutions

Guest Editor

Daochen Zhu

MDPI

Basel • Beijing • Wuhan • Barcelona • Belgrade • Novi Sad • Cluj • Manchester

Guest Editor
Daochen Zhu
School of the Environment
and Safety Engineering
Jiangsu University
Zhenjiang
China

Editorial Office
MDPI AG
Grosspeteranlage 5
4052 Basel, Switzerland

This is a reprint of the Special Issue, published open access by the journal *Toxics* (ISSN 2305-6304), freely accessible at: https://www.mdpi.com/journal/toxics/special_issues/F0623E07C7.

For citation purposes, cite each article independently as indicated on the article page online and as indicated below:

Lastname, A.A.; Lastname, B.B. Article Title. *Journal Name* **Year**, *Volume Number*, Page Range.

ISBN 978-3-7258-5763-0 (Hbk)
ISBN 978-3-7258-5764-7 (PDF)
https://doi.org/10.3390/books978-3-7258-5764-7

Contents

About the Editor

Daochen Zhu

Daochen Zhu is Executive Dean of the Institute of Biomass Energy at Jiangsu University and serves as Deputy Director of the Joint Laboratory for Artificial Intelligence Design and Biomimetic Biomass Processing, and Director of the China–Pakistan International Joint Laboratory for Synthetic Biology and Biorefinery. His research interests span microbial and enzyme discovery and industrial application, synthetic biology and environmental health, high-value utilization of agroforestry biomass, and natural product biosynthesis and metabolic regulation. He leads and participates in national and provincial research programs on biomass conversion, environmental remediation, and green manufacturing, including key research and development initiatives and international cooperation projects. He has published more than one hundred and forty SCI-indexed papers as first or corresponding author, holds multiple invention patents including an international patent, and has edited or co-edited five English-language volumes with major academic presses. Recognitions include selection to the global top two percent scientist list and several talent awards from Jiangsu Province. He also serves as Associate Editor or Guest Editor for several journals. This background supports his role as Guest Editor for the Reprint on heavy metal contamination and solutions.

Preface

This Reprint addresses the contemporary challenge of heavy metal contamination across soil, water, sediments, air, and food systems and its consequences for ecosystems and human health. Its scope spans occurrence and speciation, monitoring and exposure pathways, human health risk assessment, and laboratory to field remediation with attention to safe utilization and circular solutions. The aim is to present current evidence and practical strategies that move beyond problem diagnosis toward prevention, cleanup, and risk reduction.

Our motivation is threefold. First, heavy metals persist and bioaccumulate, creating long-lived risks that standard controls do not always resolve. Second, real-world settings often involve co-occurring stressors and complex matrices that require integrated study designs and comparable metrics. Third, decision makers need actionable data on performance, cost, and feasibility to guide investment and policy. By bringing diverse perspectives together, this Reprint highlights methods and case studies that close the gap between bench, pilot, and practice.

We thank the authors and reviewers for their careful work and the editorial team for their support. We hope this Reprint serves as a timely reference and a catalyst for collaborative progress in environmental protection and public health.

Daochen Zhu
Guest Editor

toxics

MDPI

Editorial

Heavy Metal Contamination and Its Effects on Ecosystems and Human Health: Challenges and Solutions

Daochen Zhu [1,2]

[1] International Joint Laboratory on Synthetic Biology and Biomass Biorefinery, Biofuels Institute, School of the Environment and Safety Engineering, Jiangsu University, Zhenjiang 212013, China; dczhucn@hotmail.com
[2] Jiangsu Collaborative Innovation Center of Technology and Material of Water Treatment, Suzhou University of Science and Technology, Suzhou 215009, China

1. Introduction

Heavy metals and other potentially toxic elements (PTEs) constitute a durable class of contaminants because they are non-degradable, widely emitted, and prone to accumulate across environmental compartments [1]. Across the world, the sources of these kinds of contamination are often the same—mining and smelting districts, industrial discharge, traffic-related particulates and road dust, legacy dumps, and belts of intensive agro-industrial activity [2]. This Special Issue aimed to advance an integrated understanding of how such contaminants move across compartments, how speciation and bioavailability govern the risk of contamination, and how remediation and management strategies can be designed to mitigate impacts to the environment and human health. The expansion of urban settlements shrouds this map with a low, chronic haze of inputs, as expanding industrial parks and mixed land uses discharge variable, often metal-bearing, effluents and runoff that conventional WWTPs treat imperfectly without site-specific polishing [3]. Meta analyses converge on the same pattern: mining enriches soils and river sediments; factories feed rivers through persistent wastewater discharge; and cities sustain exposure through dust that recirculates long after the smokestacks quiet down [4,5]. The consequences are familiar, pronounced "hotspot-to-urban" gradients and a slow, steady reloading of downstream sinks even as primary emissions fall. Once contaminants are released, risk is governed less by totals than by speciation and bioavailability, because organisms experience the form that is actually in a solution or at reactive surfaces. Partitioning among dissolved, colloidal, and particulate phases, together with complexation and competition at biological ligands, sets the effective dose. In aquatic systems this principle is operationalized by the Biotic Ligand Model, which underpins U.S. EPA copper criteria and is now evolving beyond equilibrium single-metal fits toward kinetic and mixture-aware formulations [6,7]. For ingestion pathways, in vitro bioaccessibility assays quantify the fraction that becomes soluble under gastrointestinal conditions, with PBET and the SBRC assay widely used in soils and dusts and the Unified BARGE Method formalized as ISO 17924 [8]. Evidence from mining-impacted soils and urban dust shows that reliance on total concentrations can result in risk being poorly estimated, and that pairing operational chemical fractionation with bioaccessibility yields more decision-relevant exposure metrics [9]. Where feasible, linking these measurements to physiologically based pharmacokinetic or toxicokinetic models, such as the IEUBK model for lead, translates an external bioaccessible dose into an internal dose and target organ metrics, tightening uncertainty in risk estimates [10].

The public health burden is substantial and measurable. For lead, no safe exposure level has been identified; the WHO and recent syntheses document enduring neurodevelopmental toxicity, and a 2019 global analysis attributed a loss of about 765 million IQ

 https://doi.org/10.3390/toxics13100837

points in children younger than five to early-life lead exposure [11,12]. Cadmium is a prototypical skeletal toxicant; meta-analyses link exposure to environmental cadmium to lower bone mineral density and higher osteoporosis risk, particularly among postmenopausal women [13–15]. Mercury remains on the WHO's list of ten chemicals of major public health concern, reflecting potent neurotoxicity with heightened vulnerability in the developing fetus. Beyond contemporary emissions, climate-amplified extremes such as floods and wildfires remobilize legacy metal stores in sediments, tailings, and floodplains, producing acute contaminant pulses superimposed on chronic exposure and often intersecting with rapid urban expansion [16]. In several regions, dietary pathways can dominate; cadmium and lead in rice supply chains are recurrent findings, which reinforces the need for mixture-aware, pathway-specific risk management.

Across this landscape, five practical questions keep returning. First, can we separate natural background contaminant levels from human inputs and state the uncertainty clearly? Second, how do we turn total concentration data into quantifications of exposure and dose by bringing in speciation, bioaccessibility, and differences in susceptibility? Third, how can we quantify and predict remobilization at the interface between sediment and water during storms, drying and rewetting, and redox shifts? Fourth, how do we protect food systems, including informal aquaculture and peri-urban agriculture, in places where diet can dominate exposure? Fifth, how do we move from single-chemical point estimates to probabilistic, mixture-aware risk assessments that directly support decisions.

This Special Issue assembles new evidence and tools across these fronts, spanning environmental monitoring and modeling, ecotoxicology and behavior, epidemiology and mechanistic mediation, food chain assessments, and intervention concepts. Together, the contributions provide a compact, methodologically diverse snapshot of where the field stands—and where it needs to go.

2. What This Special Issue Adds

River systems and integrated risk indices. One contribution synthesizes PTE contamination across rivers, combining multi-metric ecological indices (e.g., potential ecological risk, a pollutant load index, and a geo-accumulation index) with human health risk resulting from the consumption of aquatic species. Beyond mapping hotspots, the review underscores the need for harmonized metrics and comparable baselines across studies to inform management priorities. Regarding human bone health and endocrine mediation, two epidemiological studies independently link metals to impaired bone mineral density (BMD) and osteoporosis risk, while advancing why these links might arise. A large NHANES analysis reports a negative association between blood manganese and BMD and explores sex steroid hormones (including SHBG and estradiol) as potential mediators, sharpening the mechanistic lens on endocrine disruption and skeletal outcomes. A biomonitoring study in postmenopausal women identifies urinary cadmium and antimony as independent correlates of higher osteoporosis odds, highlighting co-exposures and population vulnerability. Together they illustrate how mixture-aware, mediator-informed epidemiology can move beyond associations and toward plausible pathways. On the subject of organ-specific toxicity and candidate interventions, moving from hazard to mitigation, a nanomedicine study develops a curcumin-loaded nanoemulsion to counter copper-induced neurotoxicity, improving pharmacokinetics and neuroprotection in preclinical models. While such delivery strategies are not a substitute for exposure reduction, they do foreshadow adjunctive, lower-toxicity therapeutics for metal-related disorders.

Biological variability matters, as demonstrated in a zebrafish study, where consistent behavioral phenotypes (bold vs. shy) were found to modulate sensitivity to cadmium and even to a common anthropogenic marker (caffeine), with consequences for locomotion and social behavior. This work reminds us that inter-individual variability, often treated as "noise", may systematically bias risk estimates if not explicitly designed into experiments and assessments. When it comes to food chain safety in aquatic agroecosystems, seeking data to keep pace with the expansion of nature-based and floating agriculture, a field investigation of lotus cultivated in floating lake gardens quantifies tissue-level accumulation, translocation to edible parts, and health risk indices. The results suggest preferential root accumulation and limited mobility into edible tissues, but also emphasize that chronic exposure and site conditions can shift risk profiles—underscoring the need for crop- and site-specific guidance.

For the remobilization of modeling and background levels, from concentrations to fluxes, a one-dimensional reaction-transport model tracks heavy metal desorption rates and background concentrations in cohesive sediments and reproduces estuarine observations. By coupling dissolved and particulate phases with diffusion, sedimentation, and turbulent exchange, the work offers a transferable boundary module for 3D hydrodynamic models and a pragmatic framework for distinguishing background signals from anthropogenic sources. In terms of deterministic and probabilistic risk, an integrated soil-risk assessment from a karst region finally applies both classical point estimate methods and Monte Carlo simulation, revealing that traditional approaches can underestimate carcinogenic risks (e.g., arsenic and chromium), and demonstrating how target organ toxicity dose metrics refine priority setting. This dual approach aligns assessment practice with the uncertainty and variability inherent in real-world exposures.

3. Remaining Gaps and a Forward Program

The next steps are to replace bulk concentrations with dose-relevant measures and to treat exposure as a dynamic mixture that shifts with hydrometeorological extremes. In practice, speciation and partitioning (for dissolved, colloidal, and particulate contaminants, and for complexation and competition) should be measured alongside validated oral and inhalation bioaccessibility to anchor physiologically based toxicokinetic modeling and yield internal dose estimates, rather than proxies. Human studies need to be designed up front for co-exposures (metals with co-occurring organics), effect modification (sex, age, and menopausal status), and mediating biology (hormones and inflammatory markers), with explicit causal frameworks to separate pathways and identify points where intervention alters risk. At the system scale, empirically calibrated desorption and particle–water exchange modules should be embedded within watershed and estuarine models that include hydrology and temperature, enabling event-based sampling, data assimilation, and forecasts of flood- and heat-driven remobilization, in order to clarify the line between background and anthropogenic signals.

Turning evidence into action demands a diversified portfolio and decision tools that acknowledge uncertainty. Source control and treatment, augmented by green infrastructure, remain the first-line defense; in defined subgroups, adjunctive biomedical approaches (for example, chelation when clinically indicated or low-toxicity antioxidant delivery) merit careful evaluation that considers feasibility, adherence, safety, and equity as well as efficacy. Food system safeguards should be built crop- and site-specifically for aquatic and informal agro-ecosystems, combining cultivar selection for low uptake, amendments that immobilize Cd/Pb/As, and risk communication tailored to chronic low-dose exposure. Regulators and communities need transparent, transferable approaches to estimate local background signals with uncertainty bands and to update soil and sediment standards accordingly. Probabilistic practice should become routine, reporting both point estimates and distributions with defensible priors and sensitivity analyses, and should be supported by FAIR data, versioned code, and minimal reporting checklists so that studies are reanalyzable and models portable.

4. Closing

The studies collected here advance the field on several fronts, harmonizing risk indices across media, clarifying mechanistic links to health, revealing the role of biological variability, safeguarding food chains, and equipping decision-makers with modern modeling and probabilistic tools. Taken together, they signal a shift from mapping contamination to managing risk: dose-relevant, mixture-aware and climate-attuned science coupled with interventions that regulators and communities can implement. The immediate priorities are standardized metrics and baselines, transparent background estimation, routine probabilistic reporting, and open, reproducible models that translate into measurable reductions in exposure and disease.

Data Availability Statement: No new data were created or analyzed in this study. Data sharing is not applicable to this article.

Acknowledgments: The authors would like to express their sincere gratitude to Xian Yuhang for valuable assistance in revising the manuscript and providing initial editorial support.

Conflicts of Interest: The author declares no conflicts of interest.

List of Contributions:

1. Hossain, M.M.; Jahan, I.; Dar, M.A.; Dhanavade, M.J.; Mamtaz, A.F.B.; Maxwell, S.J.; Han, S.; Zhu, D. A Review of Potentially Toxic Elements in Sediment, Water, and Aquatic Species from the River Ecosystems. *Toxics* **2025**, *13*, 26. https://doi.org/10.3390/toxics13010026.
2. Hake, G.; Mhaske, A.; Shukla, R.; Flora, S.J.S. Copper-Induced Neurodegenerative Disorders and Therapeutic Potential of Curcumin-Loaded Nanoemulsion. *Toxics* **2025**, *13*, 108. https://doi.org/10.3390/toxics13020108.
3. Santos, N.; Reis, S.; Domingues, I.; Oliveira, M. Does Personality Modulate the Sensitivity to Contaminants? A Case Study with Cadmium and Caffeine. *Toxics* **2025**, *13*, 147. https://doi.org/10.3390/toxics13030147.
4. Zhao, X.; Li, J.; Yu, J.; Shi, Y.; Tang, M. The Role of Sex Steroid Hormones in the Association Between Manganese Exposure and Bone Mineral Density: National Health and Nutrition Examination Survey 2013–2018. *Toxics* **2025**, *13*, 296. https://doi.org/10.3390/toxics13040296.
5. Elbagory, M.; Moghanm, F.S.; Mohamed, I.; El-Nahrawy, S.; Omara, A.E.-D.; Goala, M.; Kumar, P.; Mioč, B.; Andabaka, Ž.; Širić, I. Health Risk Assessment of Potentially Toxic Element Uptake by Lotus (*Nelumbo nucifera*) in Floating Lake Gardens. *Toxics* **2025**, *13*, 306. https://doi.org/10.3390/toxics13040306.
6. Gonzalez Cano, W.T.; Lonin, S.; Kim, K. Modeling Desorption Rates and Background Concentrations of Heavy Metals Using a One-Dimensional Approach. *Toxics* **2025**, *13*, 421. https://doi.org/10.3390/toxics13060421.
7. Kunioka, C.T.; de Oliveira Souza, V.C.; Rocha, B.A.; Júnior, F.B.; Belo, L.; Manso, M.C.; Carvalho, M. Association of Urinary Cadmium and Antimony with Osteoporosis Risk in Postmenopausal Brazilian Women: Insights from a 20 Metal(loid) Biomonitoring Study. *Toxics* **2025**, *13*, 489. https://doi.org/10.3390/toxics13060489.
8. Li, Q.; Li, D.; Wang, Z.; Sun, D.; Zhang, T.; Zhang, Q. Integrated Deterministic and Probabilistic Methods Reveal Heavy Metal-Induced Health Risks in Guizhou, China. *Toxics* **2025**, *13*, 515. https://doi.org/10.3390/toxics13060515.

References

1. Macklin, M.G.; Thomas, C.J.; Mudbhatkal, A.; Brewer, P.A.; Hudson-Edwards, K.A.; Lewin, J.; Scussolini, P.; Eilander, D.; Lechner, A.; Owen, J.; et al. Impacts of metal mining on river systems: A global assessment. *Science* **2023**, *381*, 1345–1350. [CrossRef] [PubMed]
2. Chen, L.; Fang, L.; Yang, X.; Luo, X.; Qiu, T.; Zeng, Y.; Huang, F.; Dong, F.; White, J.C.; Bolan, N.; et al. Sources and human health risks associated with potentially toxic elements (PTEs) in urban dust: A global perspective. *Environ. Int.* **2024**, *187*, 108708. [CrossRef] [PubMed]
3. Kato, S.; Kansha, Y. Comprehensive review of industrial wastewater treatment techniques. *Environ. Sci. Pollut. Res. Int.* **2024**, *31*, 51064–51097. [CrossRef] [PubMed]
4. Isinkaralar, O.; Isinkaralar, K.; Nguyen, T.N.T. Spatial distribution, pollution level and human health risk assessment of heavy metals in urban street dust at neighbourhood scale. *Int. J. Biometeorol.* **2024**, *68*, 2055–2067. [CrossRef] [PubMed]
5. Yuan, X.; Xue, N.; Han, Z. A meta-analysis of heavy metals pollution in farmland and urban soils in China over the past 20 years. *J. Environ. Sci.* **2021**, *101*, 217–226. [CrossRef] [PubMed]
6. U.S. Environmental Protection Agency. *Aquatic Life Ambient Freshwater Quality Criteria—Copper 2007 Revision*; U.S. EPA: Washington, DC, USA, 2007.
7. U.S. Environmental Protection Agency. *Copper Biotic Ligand Model*; U.S. EPA: Washington, DC, USA, 2007.
8. *ISO 17924:2018*; Soil Quality—Procedure for the Estimation of the Human Bioaccessibility/Bioavailability of Metals in Soil. International Organization for Standardization: Geneva, Switzerland, 2018.
9. Soltani, N.; Keshavarzi, B.; Moore, F.; Cave, M.; Sorooshian, A.; Mahmoudi, M.R.; Ahmadi, M.R.; Golshani, R. In vitro bioaccessibility, phase partitioning, and health risk of potentially toxic elements in dust of an iron mining and industrial complex. *Ecotoxicol. Environ. Saf.* **2021**, *212*, 111972. [CrossRef] [PubMed]
10. U.S. Environmental Protection Agency. *User's Guide for the Integrated Exposure Uptake Biokinetic (IEUBK) Model for Lead in Children*; U.S. EPA: Washington, DC, USA, 2021.
11. Larsen, B.; Sánchez-Triana, E. Global health burden and cost of lead exposure in children and adults: A health impact and economic modelling analysis. *Lancet Planet Health* **2023**, *7*, e831–e840. [CrossRef] [PubMed]
12. Kunioka, C.T.; Manso, M.C.; Carvalho, M. Association between Environmental Cadmium Exposure and Osteoporosis Risk in Postmenopausal Women: A Systematic Review and Meta-Analysis. *Int. J. Environ. Res. Public Health* **2022**, *20*, 485. [CrossRef] [PubMed]
13. Lei, Y.; Guo, M.; Xie, J.; Liu, X.; Li, X.; Wang, H.; Xu, Y.; Zheng, D. Relationship between blood cadmium levels and bone mineral density in adults: A cross-sectional study. *Front. Endocrinol.* **2024**, *15*, 1354577. [CrossRef] [PubMed]
14. Crawford, S.E.; Brinkmann, M.; Ouellet, J.D.; Lehmkuhl, F.; Reicherter, K.; Schwarzbauer, J.; Bellanova, P.; Letmathe, P.; Blank, L.M.; Weber, R.; et al. Remobilization of pollutants during extreme flood events poses severe risks to human and environmental health. *J. Hazard. Mater.* **2022**, *421*, 126691. [CrossRef] [PubMed]
15. World Health Organization. *Mercury and Health, Fact Sheet*; World Health Organization (WHO): Geneva, Switzerland, 2024.
16. Murphy, S.F.; McCleskey, R.B.; Martin, D.A.; Holloway, J.M.; Writer, J.H. Wildfire-driven changes in hydrology mobilize arsenic and metals from legacy mine waste. *Sci. Total. Environ.* **2020**, *743*, 140635. [CrossRef] [PubMed]

toxics

MDPI

Review

A Review of Potentially Toxic Elements in Sediment, Water, and Aquatic Species from the River Ecosystems

Md Muzammel Hossain [1,2,3], Iffat Jahan [3,4], Mudasir A. Dar [1], Maruti J. Dhanavade [5], Al Fattah Bin Mamtaz [6], Stephen J. Maxwell [7], Song Han [1,2] and Daochen Zhu [1,2,*]

1 Biofuels Institute, School of Environment and Safety Engineering, Jiangsu University, Zhenjiang 212013, China; 1000006431@ujs.edu.cn (M.M.H.); muddar7@gmail.com (M.A.D.); hansong@ujs.edu.cn (S.H.)
2 Jiangsu Collaborative Innovation Center of Technology and Material of Water Treatment, Suzhou University of Science and Technology, Suzhou 215009, China
3 Biodiversity Conservation and Fisheries Research Center, Dhaka 1207, Bangladesh; ch18032@mbstu.ac.bd
4 Department of Chemistry, Mawlana Bhashani Science and Technology University, Santosh 1902, Bangladesh
5 Department of Microbiology, Bharati Vidyapeeth's Dr Patangrao Kadam Mahavidyalaya, Sangli 416416, India; marutijd@gmail.com
6 Institute of Agribusiness & Development Studies, Bangladesh Agricultural University, Mymensingh 2202, Bangladesh
7 College of Science and Engineering, James Cook University, Cairns, QLD 4878, Australia; stephen.maxwell@my.jcu.edu.au
* Correspondence: dczhucn@ujs.edu.cn

Abstract: There is concern over potential toxic elements (PTEs) impacting river ecosystems due to human and industrial activities. The river's water, sediment, and aquatic life are all severely affected by the release of chemical and urban waste. PTE concentrations in sediment, water, and aquatic species from river ecosystems are reported in this review. Among the PTEs, chromium (Cr), cadmium (Cd), lead (Pb), and nickel (Ni) revealed high pollution levels in water and aquatic species (fish and shellfish) at many rivers. The Karnaphuli, Ganga, and Lee rivers have high levels of Pb and Cd contamination, while the Buriganga and Korotoa rivers' water had notable Ni contamination. A number of rivers with PTEs showed ecological risk as a consequence of the sediment's potential ecological risk (PER), the pollutant load index (PLI), and the geoaccumulation index (Igeo). A comprehensive study suggests elevated PLI values in river sediments, indicating significant pollution levels, particularly in the Buriganga River sediment, marked by high Igeo values. The PER of the Shitalakshya and Buriganga rivers was marked as very high risk, with an $E^i_r > 320$, while the Dhaleshwari and Khiru rivers showed 'high risk', with 160 $= E^i_r < 320$. It was found that fish and shellfish from the Buriganga, Turag, and Swat rivers have a high concentration of Cr. PTE pollution across several river sites could pose health toxicity risks to humans through the consumption of aquatic species. The CR value shows the carcinogenic risk to human health from eating fish and shellfish, whereas an HI value > 1 suggests no carcinogenic risk. The occurrence of other PTEs, including manganese (Mn), arsenic (As), and nickel (Ni), significantly increases the ecological risk and concerns to aquatic life and human health. This study emphasises the importance of PTE toxicity risk and continuous monitoring for the sustainability of river ecosystems.

Keywords: PTE; river; risk assessment; anthropological activities; monitoring

1. Introduction

The river is the main source of nutrients for life. In recent decades, the proximity of river basins to urban and rural areas has exacerbated the issue of potential toxic element

https://doi.org/10.3390/toxics13010026

(PTE) contamination in river sediments, water, and aquatic species [1–3]. Urbanization, industrialization, and anthropological activities have led to the accumulation of PTEs, posing a momentous threat to aquatic life and river ecosystem health. The Bangladesh Economic Review [4] states that industrial activities have increased in different urban areas. In urban settings, many rivers close to urban centres face heightened risks of contamination from sediment pollution, water pollution, household waste, industrial waste, and chemical discharges [5,6]. Riverbank ecosystems are vital components of aquatic habitats, supporting diverse forms of life, including sediment, soil, flora, and fauna. These ecosystems perform a crucial role in maintaining environmental quality and ecological balance. Particularly, they have experienced a rapid increase in PTE pollution within their river systems due to unregulated industrial growth, inadequate waste management practices, and the discharge of hazardous chemicals [7–9]. The sources of PTE contamination are diverse, encompassing industrial waste disposal, mining, smelting, and the improper handling of wastewater and chemical waste [10]. The uncontrolled discharge of vehicle emissions and sewage sludges into rivers is a major contributing factor to the contamination of the aquatic environment [11]. Large volumes of PTEs containing effluent from industries are typically dumped into adjacent water bodies, endangering both the sustainability of the environment and public health [12–14]. Based on Theofanis et al. [15], the consequences of PTE contamination extend beyond the immediate aquatic environment. Sediment enriched with PTE pollution becomes part of the aquatic food chain, leading to the accumulation of these toxic elements in living organisms through bio-magnification. The amount of PTEs present in the sediment serves as an important indicator of this equilibrium since it shows how different human activities have affected the health of river ecology. The sediment ecology is threatened by contamination by PTEs [16]. Because of their bioaccumulative nature, toxicity, and environmental persistence, they are regarded as environmental contaminants [17,18]. Above-threshold concentrations of PTEs may accumulate in the biota of riverine ecosystems and have detrimental impacts on both animals and people [19,20]. The potential outcomes of this contamination range from negative impacts on organisms to potential declines in species diversity, abundance, and toxicity risks to human health. Despite the increasing recognition of PTEs, pollution in river sediments worldwide remains a significant gap in ecological research, particularly in the context of developed countries. According to Bashar & Fung [21], the leather and textile industries are potential sources of Pb, Cd, and Ni, which are the principal causes of river sediment pollution. Sediment ecology is negatively affected by Cd and Pb due to anthropological activity. The lack of comprehensive studies addressing PTE contamination across different components of river ecosystems is a critical concern for both environmental and human health due to excessive contamination, and the bio-magnification and toxicity of these pollutants. PTE contamination in sediments may pose a serious ecological concern to urban river ecosystems [22–24], although few studies have examined the pollution level of PTEs in river sediment, water, and aquatic species, raising concern over the PTE toxicity risk in ecosystems and for human health. There is very little research on the PTE contamination risk in riverine environments. Therefore, as an initial effort to bridge the research gap, the ecological risks associated with PTE pollution in river sediments, water, and aquatic organisms are comprehensively assessed. River sediment, water, and aquatic species risk may be understood by taking into account the origins, distribution, graphical representation of PTE concentration, and impact of PTE pollution.

Source of Pollution

River sites in both urban and rural environments are contaminated by a variety of factors, such as construction sites, decomposition of organic matter, logging operations, soil erosion, domestic and industrial chemicals and waste, farming practices, agricultural drainage, and storm-water runoff. River ecosystems face difficult problems as a result of the interaction of various elements. Approximately 15,666 small and 3639 big manufacturing enterprises have been established in Bangladesh, according to the BBS [25]. Proshad et al. [26] reported that a wide range of industries have emerged as a result of rapid industrialization. A glaring concern arises from the discover industries discharging untreated waste directly into the river, raising significant health concerns for the environment and public. Regrettably, insufficient monitoring and inadequate responses have catalysed an escalation in environmental degradation. Yet, the ominous cloud of pollution looms over several river ecosystems, stemming from various sources, including floods, soil erosion, and the discharge of PTEs, chemicals, herbicides, and insecticides, as well as solid and liquid industrial wastes. In the district of Gazipur, Dhaka a staggering 2220 factories operate, with 1222 dedicated to ready-made clothing production. Surprisingly, only 1% of these factories have functioning treatment plants, exposing a vast majority of them to the risk of unregulated waste discharge [27]. In comparison to other factories, the Department of Environment (2016) reported that enforcement activities were higher in fabric-washing factories (38%) and fabric-dying factories (30%). The perilous scale of waste disposal becomes even more apparent when considering the report by Kamruzzaman and Sakib [28], which highlights that around 350 metric tons of harmful garbage are dumped into rivers. Oil refineries add to the burden, with solid waste amounting to 4 tonnes and liquid waste reaching 0.61 million cubic meters annually. The textile dyeing and tanning industries alone contribute significantly, annually generating 113.72 tonnes of solid waste and a staggering 26,250 tons of liquid waste, both of which find their way into river systems, compromising water quality [29]. Soil erosion and mountain cutting stand as significant contributors, accounting for approximately 3% of overall environmental degradation [29]. The DoE [30] attributes approximately 10% of pollution to a combination of factors, encompassing activities such as logging, forest fires, rainwater runoff, and the decomposition of plants and animals. Water pollution is largely caused by these complex biological interdependencies, riverside farming, and agricultural runoff from agricultural regions. Nearshore aquatic habitats provide the most vital ecological functions, from food provisioning to support for human activities, as highlighted by Arikibe and Prasad [31]. This review reveals a concerning interplay between pollution activities and PTE concentrations across distinct river regions. By shedding light on these intricate dynamics, we emphasise the need for targeted interventions and enhanced monitoring strategies to mitigate the escalating ecological risks and human health risks.

2. Screening and Systematic Approach

The concentration of PTE in sediment, water, and aquatic organisms, as well as concerns about excessive PTE contamination, are the study's objectives. To achieve our goals, we have chosen and employed a systematic approach, division-wise, utilising information from the Web of Science and Scopus databases, employing relevant search terms such as "river sediment", "river water", "aquatic species", "heavy metal", "PTEs", and "element". For geographical limitation, the selection process involved critically evaluating and choosing the most suitable original research papers in English from the related journal groups such as Elsevier, Springer, Taylor & Francis, PLOS ONE, and others from 2010 to 2024. All the collected information underwent rigorous screening to establish a well-structured portfolio, laying the foundation for the final outcomes of this study and providing directions for future investigations. An outline of the general approach that shows the key procedures and analyses performed in the current investigation is shown in Figure 1. The sediment data visualisation method follows the USEPA standard risk model [32]. We consider the major points for data assembly and analysis. Sediment, water, and aquatic species sample data analysis followed standard protocols [33–35].

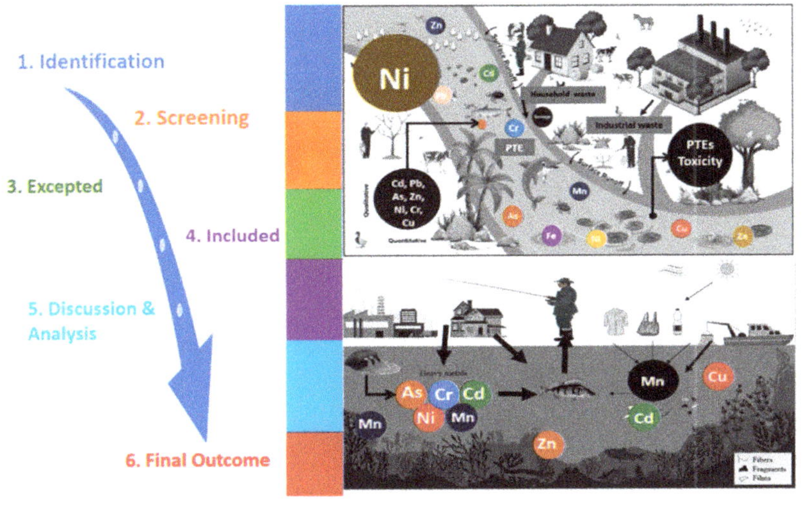

Figure 1. Flow chart of the overall methodology indicating major steps.

In Table 1, several PTE indices, indexing factors, and recommendations are described along with their respective contamination standards. Assessment of the pollution risks from PTEs in river sediment, water, and aquatic species are compared based on following the scientific method for monitoring and safety in river ecology. This treasure trove of information subsequently underwent meticulous scrutiny against established Sediment Quality Guidelines (SQGs), affording a comprehensive evaluation of the ecological risks posed by potential toxic element (PTE) accumulation in these vital aquatic systems.

Table 1. PTE indices, indexing factors, description, and objectives.

Principal Indexing Factor	Description and Objectives
$C^i_f = C^i / C^i_n$ $C_d = \sum C^i_f\,(n, i = 1)$ $\text{PLI} = (\text{Cf1} \times \text{Cf2} \times \text{Cf3} \times \text{Cf n})^{1/n}$ $I_{geo} = \text{Log}_2\,(\frac{\text{Cn}}{1.5 \times \text{Bn}})$ $E^i_r = T^i_r \times C^i_f$ $\text{PER} = \sum E^i_r\,(\text{i} = 1)$ $THQ = \frac{CM}{MRL}$ $HI = \sum THQ(\text{x})$ $TCR = \frac{EF \times ED \times FIR \times Cf \times CM \times CPSo}{WAB \times TAc} \times 10^{-3}$	PTEs is potential toxic elements. In the equation, C^i_f is a contamination factor; C^i is the quantified value of PTE in sediments; and the elemental reference value for the same metal follows Taylor [36]; and C^i_n is the background concentration following Hilton et al. [37] and Karadede and Unlu [38]. C_d is the degree of contamination; PLI is the Pollution Load Index, following Suresh et al. [39]. I_{geo} is the geoaccumulation index following Müller's [40] technique. C_n is the metal concentration in sediment (n); B_n is the metal (n)'s geochemical background value; and the factor 1.5 is the possible variation in background data to lithogenic impacts, following Rabee et al. [41]. E^i_r is the potential ecological risk index for an individual element. T^i_r is the biological toxic factor for individual elements and is 5, 30, 2, 5, 1, and 6 for Pb, Cd, Cr, Cu, Zn, and Ni, respectively, following Hakanson [42]. PER is the potential ecological risk index, follows Guo et al. [43] and Hossain et al.'s [22] technique. Non-carcinogenic and carcinogenic: THQ, Target hazard quotient; HI, Hazard Index; TCR, Target Cancer Risk.

Table 1. *Cont.*

Principal Indexing Factor	Description and Objectives
Contamination status indication for sediment	Recommendation level: Reference value (mg/kg) Contamination as I_{geo}: $I_{geo} \leq 0$, no contamination; $0 \leq I_{geo} \leq 1$, no contamination to moderately contamination; $1 \leq I_{geo} \leq 2$, moderately contaminated; $2 \leq I_{geo} \leq 3$, moderately to heavily contaminated; $3 \leq I_{geo} \leq 4$, heavily contaminated; $4 \leq I_{geo} \leq 5$, heavily to extremely contaminated; $5 < I_{geo}$, extremely contaminated. Contamination as CF: $CF < 1$, low; $1 < CF < 3$, moderate; $3 < CF < 6$, considerable; $CF < 6$, very high. Contamination as PLI: $PLI = 0$, indicates excellence; $PLI = 1$, this level is contaminated; $PLI > 1$, the quality is gradually declines.
Contamination Status for water and aquatic species	Recommendation level (mg/kg) Reference value (mg/kg) Toxic level (mg/kg)

3. PTE Concentration and Distribution

The concentration and distribution of PTEs in river sediments, water, and aquatic species show regional variations formed by the particular interaction of geographical variables and anthropological activity and influenced by the specific river site and the amount of PTE pollution. A notable reference enriches our understanding, providing the lowest effect levels (LELs) for diverse PTEs in sediment. For instance, Pb exhibits an LEL at 600 mg/kg, Zn at 31 mg/kg, Cu at 120 mg/kg, Cd at 16 mg/kg, Cr at 6 mg/kg, and Ni at 26 mg/kg, among others.

3.1. PTEs in Sediments

3.1.1. Lead (Pb)

Pb is highly toxic to aquatic organisms within contaminated sediments in the ecosystem. High Pb content was found in the Korotoa River (64.67 mg/kg), Rupsha River (62.40 mg/kg), and Buriganga River (477.87 mg/kg), and all values were over background limits (Table 2). Among the rivers in Bangladesh, the Buriganga River sediment exhibits the highest concentration, followed by the Korotoa, Rupsha, Bangshi, Karnaphuli, Turag, Shitalakhya, Dhaleshwari, Meghna, Brahmaputra, and Louhajang rivers. Most river sites surpass the background levels of FAO and SEPAC for Pb, except the Brahmaputra River and Louhajang River sites. Worldwide river sites: other researchers have found Pb pollution in sediment such as in the Yellow River [44] and Xiangjiang River, China [45,46], Gomti River, India [47], Gorges River, Australia [48], Louro River, Spain [49], Symsarna River, Poland [50], and Elbe River, Germany [51] (Table 3). Recently, Proshad et al. [52] reported a high concentration (64.67 mg/kg) of Pb in sediment at the Korotoa River. The Pb contamination in the Buriganga River area stems from the improper disposal of residential and commercial waste, as well as sewage sludge, which has consequently contaminated the sedimentary environment. The highest relative abundance of Pb was found in the Karnaphuli River (Figure 2), with a similar abundance was found in the Symsarna River, Poland (Figure 3). Pb pollution occurred in the river Karnaphuli due to industrial discharge. Specifically, regions such as Suthrapur, Lalbag, and Shyampur within the Buriganga River system demonstrate significant Pb contamination. The FAO recommend Pb concentration of 5 mg/kg in sediment, respectively. The USEPA reported toxicity range from 21 mg/kg to 20 mg/kg for Pb, with the minimum impact of Pb content (31 mg/kg) [53], which raises serious concerns regarding sediment pollution.

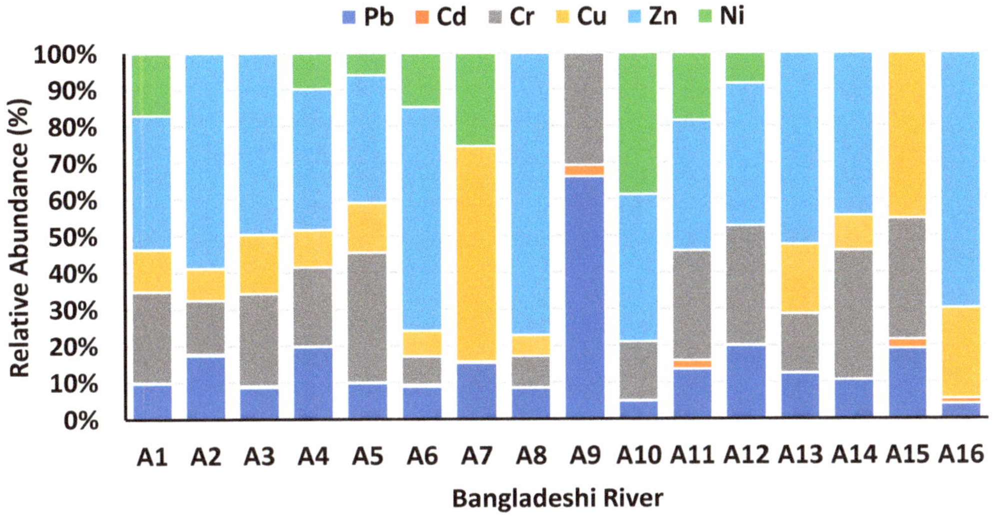

Figure 2. Relative abundance (%) of PTEs in sediment of Bangladeshi river. (Here, A5: Buriganga River; A1: Korotoa River; A9: Karnaphuli River; A2: Meghna River; A3: Shitalakshya River; A4: Rupsha River; A6: Brahmaputra River; A7: Louhajang River; A8: Halda River; A10: Meghna River; A11: Shitalakshya River; A12: Bangshi River; A13: Turag River; A14: Padma River; A15: Dhaleshwari River; A16: Khiru River).

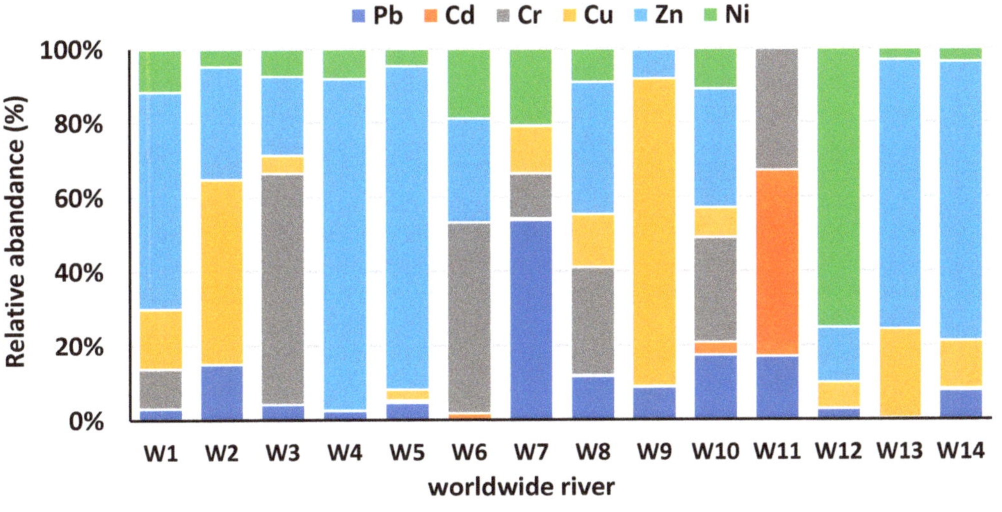

Figure 3. Relative abundance (%) of PTEs in sediment of rivers worldwide. (Here, W7: Symsarna River, Poland; W14: Elbe River, Germany, W10: Ganga River, India; W9: Lubumbashi River, Congo; W11: Okumeshi River, Nigeria; W8: Yellow River, China; W6: Pra River, Ghana; W3: Atoyac River, Mexico; W1: SomesuMic River, Romania; W2: Barma River, Malaysia; W4: Saigon River, Vietnam; W5: Lee River, England; W12: Buyukmelen River, Turkey; W13: Liffey River, Ireland).

Industrial discharges, urban runoff, and atmospheric deposition are common sources of Pb in sediment. Pb toxicity in sediment is a significant environmental concern as it can have adverse effects on aquatic ecosystems and pose risks to human health.

3.1.2. Cadmium (Cd)

Cd is a toxic element, and high levels of Cd were reported in the areas of the Buriganga (5.86 mg/kg), Shitalakshya (5.01 mg/kg), Dhaleshwari (2.08 mg/kg), Khiru (2.05 mg/kg), and Karnaphuli (2.01 mg/kg) rivers in Bangladesh, exceeding the background values of SEPAC (Table 2). The Buriganga River site showed the highest Cd contamination. The Shitalakshya, Dhaleshwari, Khiru, and Karnaphuli River sites's tapestry tells a fascinating story for Bangladesh. For Cd in sediment, scientifically the lowest impact threshold is 0.6 mg/kg (USEPA).The Hazaribag and Lalbag areas in the Buriganga River basin are conspicuous instances of high levels of Cd contamination, necessitating prompt action to alleviate any risks. Similarly, worldwide, high levels of Cd pollution have been discovered in sediment at the Ganga River, India [54]; Nile River, Egypt [55]; Gardon of Ales River, France [56]; and Lubumbashi River, Congo [57] (Table 3). The USEPA [58] reported Cd toxicity value is 1 mg/kg and guideline limit is 0.61 mg/kg in sediment. A high relative abundance of Cd was reported in the Karnaphuli River in Bangladesh [59] (Figure 2), whereas Raphael et al. [60] identified a comparable abundance at the Okumeshi River in Nigeria (Figure 3). Comparable levels of Cd pollution have been documented around the world, indicating the worldwide reach of the problem and emphasizing the need for efficient mitigation techniques.

3.1.3. Chromium (Cr)

The ecological effects of Cr contamination in sediment ecosystems depend on the oxidation state and concentration of the metal. The most significant amount of Cr in the winter (summer) season is recorded at 158.37 mg/kg (140.53 mg/kg), which is the second highest level of metals for the Buriganga River. The Turag River has the third highest metal levels during the winter (summer) season, with a value of 120.15–30.27 (118.2–28.15) mg/kg. The highest Cr concentration was found in the Buriganga River site, followed by the Korotoa, Bangshi, Rupsha, Turag, Padma, Shitalakhya, Dhaleshwari, Karnaphuli, and Meghna rivers (Table 2). Recently, Proshad et al. [52] found the highest amount of Cr (165.84 mg/kg) in sediment from the river Korotoa. Mohiuddin et al. [61] reported high mean concentrations of Cr (173.4 mg/kg and 709.41 mg/kg) in Buriganga River sites, exceeding the guideline value.The Bangshi River sediment sample is noteworthy due to its Cr content of 33%. Important Cr pollution is found in areas of the Buriganga River system, including the Hazaribag and Lalbag sites. Outside the nation, numerous researchers have discovered Cr pollution in sediment of various rivers, including the Yellow River, China [62], the Pra River, Ghana [63], and the Atoyac River, Mexico (Table 3), which is a global concern for Cr pollution. A high relative abundance of Cr in sediment was found at the Buriganga River in Bangladesh (Figure 2), whereas Rodrigue-Zespinosa et al. [64] discovered a comparable abundance at the Atoyac River in Mexico (Figure 3). The USEPA established a toxicity reference value of 8.10 mg/kg, which raises concerns about Cr pollution in river sediment. Instead, SEPAC [65] reported a background value of 66.80 mg/kg for Cr. Turekian and Wedepohl [66] state the average shale value of Cr is 90 mg/kg whereas the WHO and the USEPA have recommended standards for sedimentary Cr values of 25 and 26 mg/kg, respectively. High levels of Cr pollution indicate a risk for ecology and are of rising concern globally.

Table 2. The mean concentration of potential toxic elements (PTEs) (mg/kg) in sediment, pollution sources, and analytical processes are discussed from Bangladesh.

Time	River Name	River study Code	City Code	Pollution Source	Analytical Method	Pb	Cd	Cr	Cu	Zn	Ni
2022	Korotoa River	A1	Bo	AI	ICP-MS	64.67	1.49	165.84	76	243.68	114.13
2021	Meghna River	A2	No	AFA	AAS	12.48	0.28	10.59	6.22	42.41	
2020	Shitalakshya River	A3	Na	IA	FAAS and GFAAS	13.16	0.64	38.39	24.6	75.48	
2020	Rupsha River	A4	K	AI	AAS	62.4	0.56	67.72	31.95	121.35	31.34
2019	Buriganga River	A5	D	IC	ICP-MS	11.405	0.23	41.45	15.93	40.71	7.14
2019	Brahmaputra River	A6	Na	TCDI	FAAS	7.6	0.48	6.6	6.2	52.7	12.8
2019	Louhajang River	A7	T	AI	ICP-MS	4.597	0.083	9. 21	17.727		7.676
2017	Halda River	A8	C	AI	AAS	8.8	0.04	8.84	5.9	79.58	
2016	Karnaphuli River	A9	C	IA	AAS, GF-AAS	43.69	2.01	20.3			
2015	Buriganga River	R11	D	IC	AAS	31.4	1.5	173.4	344.2	481.8	153.3
2015	Meghna River	A10	Nar	IA	AAS	9.47	0.23	31.74		79.02	76.12
2015	Korotoa River	R12	Bo	AI	ICP-MS	58	1.2	109	76		95
2014	Shitalakshya River	A11	Nar	IA	AAS	28.36	5.01	63.22		75	39.22
2014	Bangshi River	A12	T	AA	AAS	59.99	0.61	98.1		117.15	25.67
2013	Turag River	A13	D	AI	AAS	32.78	0.28	43.02	50.4	139.48	
2013	Padma River	A14	D	AI	AAS	11.7		38.91	10.64	49.16	
2012	Dhaleshwari River	A15	S	AI	FAAS	15.79	2.08	27.39	37.45		
2012	Khiru River	A16	M	AA	AAS	5.6	2.05		34.7	97.77	
2011	Buriganga River	R13	D	IC	ICP-MS	477.85	5.86	709.41	224.55	958.15	137.35
2011	Buriganga River	R14	D	IC	AAS	79.8	0.8	101.2	184.4	502.3	
2010	Buriganga River	R15	D	IC	ICP-MS	69.75	3.25	174.53	30.35		200.45
	Toxicity Ref. Value (USEPA)	R16				21.00	1	8.10	28.00	68.00	
	Background Value (SEPAC)	R17				21.90	0.08	66.80	25.50	69.60	33.80
	Average shale Value	R18					0.30	90.00		95.00	

Note: Bo: Bogra, D: Dhaka, No: Noakhali, Na: Narsingdi, Nar: Narayanganj, K: Khulna, T: Tangail, C: Chittagong, S: Savar, M: Mymensingh; ICP-MS: inductively coupled plasma mass spectrometry, AAS: atomic absorption spectroscopy, FAAS: flame atomic absorption spectroscopy, GFAAS: graphite furnace atomic absorption spectroscopy; IC: industrial and commercial; AA: agricultural area; AI: agricultural and industrial; IA: industrial area; TCDI: textile craft and dyeing industries; AFA: agricultural and fishing area. Here, A5 [33], A1 [52], R16 [58], A9 [59], R11 [61], R17 [65], R18 [66], A2 [67], A3 [68], A4 [69], A6 [70], A7 [71], A8 [72], A10 [73], R12 [74], A11 [75], A12 [76], A13 [77], A14 [78], A15 [79], A16 [80], R13 [81], R14 [82], R15 [83].

3.1.4. Copper (Cu)

Globally, diverse Cu concentrations have been recorded, emphasizing the importance of comprehensive monitoring and effective management strategies.Siddique et al. [67] reported Cu concentration in sediment. The maximum concentration of Cu (76 mg/kg) was found in sediment of the river Korotoa. Due to leakage, runoff from nearby unhealthy farms, and the discharge of industrial and municipal trash into the waterway, the Turag River had a higher level of Cu than the Buriganga River. In order of high Cu concentration, the Buriganga River is followed by the Korotoa, Turag, Dhaleshwari, Khiru, Rupsha, Shitalakhya [68], and Louhajang rivers that reveal the background value in Bangladesh (Table 2). Based on USEPA, toxicity value is 28 mg/kg whereas Rupsha river cross the limit [69]. These levels can cause stress and have an impact on aquatic life. Among the rivers, the highest mean concentration (344.20 mg/kg) of Cu in the Buriganga River site exceeded the FAO limit and the background levels of SEPAC. Cu pollution in sediment has been reported worldwide, such as in the Yellow River in China, SomesuMic River, Romani, Barma River, Malaysia, and Liffey River, Ireland(Table 3). A relatively high abundance of Cu was reported at the Louhajang River in Bangladesh (Figure 2), whereas a comparable abundance was found at the Lubumbashi river, Congo (Figure 3). The lowest effect threshold level is 16 mg/kg and in this case Brahmaputra River showed safety zone [70]. Interestingly, notable Cu levels were shown at the Louhajang [71] and Dhaleshwari river sites.

Table 3. Potential toxic elements (PTEs) in sediment from different rivers. worldwide.

			Sediment (mg/kg)					
River Name	**Country**	**Study Code**	**Pb**	**Cd**	**Cr**	**Cu**	**Zn**	**Ni**
Symsarna River	Poland	W7	87.32	0.69	19.76	20.63	0.13	33.88
Elbe River	Germany	W14	122	7.3		206	1190	58
Ganga River	India	W10	151.85	30.01	247.05	70.7	278.61	97.1
Lubumbashi River	Congo	W9	1549	42.9		14,822	1415	55.4
Okumeshi River	Nigeria	W11	0.45	1.32	0.87			
Yellow River	China	W8	24.6	0.3	61.3	30.3	74.6	19.3
Pra River	Ghana	W6		7.27	216.7		118.32	79.9
Atoyac River	Mexico	W3	12		182	14	62	22
SomesuMic River	Romania	W1	12.27	0.35	43.15	65.56	236.82	47.69
Barma River	Malaysia	W2	123			410	250	40
Saigon River	Vietnam	W4	2	0.07			75	6.93
Lee River	England	W5	50	5.64		32.6	946	51.1
Buyukmelen River	Turkey	W12	12.1	0.12		30.6	63.7	323
Liffey River	Ireland	W13		3.25		220	666	29

Note: Here, river study code is W7 [50], W14 [51], W10 [54], W9 [57], W11 [60], W8 [62], W6 [63], W3 [65], W1 [84], W2 [85], W4 [86], W5 [87], W12 [88], W13 [89].

3.1.5. Zinc (Zn)

The amount of Zn in the sediment indicates the level of pollution in the river ecosystem. While Zn is a crucial trace element for many species, high Zn concentrations can impact sediment ecosystems.The highest mean concentration (958.15 mg/kg and 502.3 mg/kg) in the Buriganga River site surpassed the background levels of SEPAC (Table 2). The Zn toxicity value is 68.00 mg/kg for sediment which is exceeded in the Halda River [72], Meghna River [73]. Comparisons with global Zn concentrations highlight the need for a comprehensive approach to mitigate Zn pollution. Islam et al. [74] reported PTE pollution in the urban river area. The Buriganga River continued a pattern of high Zn content, followed by the Korotoa, Turag, Rupsha, Bangshi [76], Khiru, Shitalakhya [75], Halda, and

Meghna rivers in Bangladesh. A high Zn concentration (139.48 mg/kg) was discovered at the Turag River site [77] and Padma river site showed safe zone [78]. Ahmed et al. [79] reported PTE pollution in the Dhaleshwari river site. In sediment, the Khiru [80] and Buriganga [81,82] rivers showed high relative abundance (%) of Zn (Figure 2), whereas identified comparable relative abundance in the Saigon River, Vietnam (Figure 3). More patterns were identified in the Hazaribag and Lalbag regions of the Buriganga River system. The USEPA reported average shale value is 95 mg/kg, which raises concerns regarding potential Zn contamination. An increasing number of people worldwide are becoming concerned about Zn contamination that has been found in sediment at the Elbe River, Germany, the Gardon of the Ales River in France, the SomesuMic River, Romania, Lee River, England and Liffey River, Ireland (Table 3).

3.1.6. Nickel (Ni)

Understanding and monitoring Ni contamination in sediment and aquatic environments is crucial for evaluating its potential environmental impacts and implementing appropriate remediation measures to safeguard the health of these ecosystems. A maximum mean value (200.45 mg/kg) of Ni was shown in the Buriganga River site [83], while the Louhajang River site exhibited the lowest concentration (7.676 mg/kg) (Table 2). Ganges river [90] site showed the water and sediment pollution. High concentrations of Ni were observed in the Buriganga, Korotoa, Meghna, Shitalakshya, Rupsha, Bangshi, Brahmaputra, and Louhajang river sites in Bangladesh, exceeding the guideline value. The Turag River had the highest level of Ni (95.1 mg/kg), which is higher than in the Buriganga River. Sediment ecology is becoming more of a concern globally due to the discovery of Ni contamination in sediment at the SomesuMic River, Romania (47.69 mg/kg) [84], Barma River, Malaysia (40 mg/kg) [85]. Recently, the highest level of Ni (114.13 mg/kg) found in sediment along the river Korotoa (Table 2). Remarkable Ni concentrations were found in the Buriganga and Shitalakshya River sites, but the Hazaribag and Lalbag areas of the Buriganga River system had higher Ni levels whereas the Saigon River, Vietnam showed the safety level of Ni concentration [86]. Relatively high Ni concentration discovered in the Lee River, England [87], Meghna River in Bangladesh (Figure 2), whereas Pehlivan [88] found a similar richness at the Buyukmelen River in Turkey (Figure 3). The USEPA and WHO suggested guideline values are 16 mg/kg and 20 mg/kg, respectively, for Ni concentrations in sediment. The dangerous limit for Ni contamination is 16 mg/kg, which is the lowest effective amount of Ni [53]. Also, a high level of Ni was reported at the Ganga River, India, Nile River, Egypt (112 mg/kg), Pra River, Ghana, and Liffey River, Ireland [89] (Table 3).Chronic exposure to Ni leads to unhealthy benthic communities, favouring species that are more tolerant to Ni contamination, altering species composition and disrupting ecosystem dynamics. Even changes in microbial communities affect nutrient cycling, sediment processes, and overall ecosystem health.

3.2. Pollution Load Index (PLI)

The sediment pollution status is represented in river areas through an in-depth analysis of PTE concentration. Various PTE results showed novel impacts as an ecological risk when using an integrated potential risk assessment for all metals from river basins. The study utilised the Pollution Load Index (PLI) as an indicator of sediment quality concerning PTEs. The PLI was calculated based on contamination factors for each metal and their corresponding background values. Yi et al. [91] reported PLI in the Yangtze River, China. Similarly PLI value found in the Buriganga and Turag river during the wintertime period ranged from 0.56 to 0.33 and 1.06 to 0.35, respectively, whereas the values for Buriganga and Turag in the summertime period were 0.51 to 0.29 and 1.006 to 0.35, respectively. PLI values exceeding 1 indicated pollution, and values below 1 indicated uncontaminated sediment. The Buriganga, Korotoa, Turag, Rupsha, Shitalakhya, Bangshi, Khiru, and Dhaleshwari rivers in Bangladesh exhibit Pollution Load Indices exceeding the permissible limit of 1, with the PLI range spanning from 0.21 to 9.34 within the research (Figure 4).

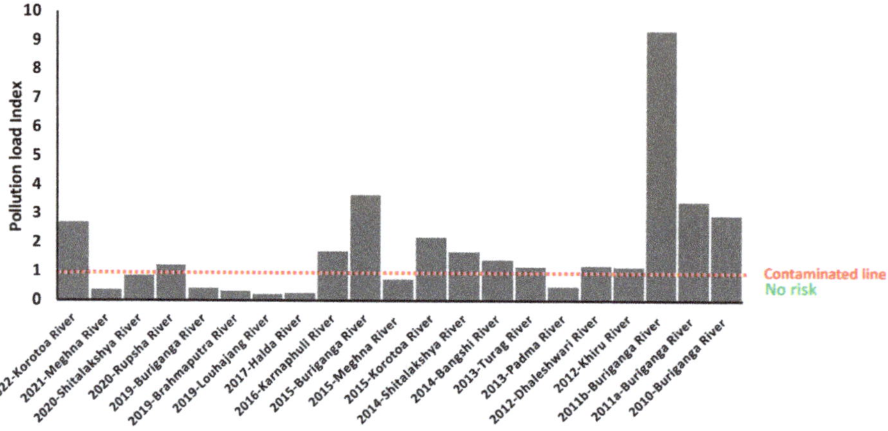

Figure 4. Pollution Load Index (PLI) is represented for different rivers and times.

The rivers examined displayed PLI values surpassing 1, indicating pollution, with the Buriganga and Shitalakshya rivers exhibiting considerable risk. The PLI value indicated that the Buriganga River sediment was in better condition in 2019 despite being severely contaminated in 2010 and 2011. The significantly raised PLI value of the Korotoa River sediment in 2022 is causing considerable concern for bottom-feeder aquatic species and river ecology. PLI value between 1.64 and 3.89 indicates that the whole stretch of the Ganga River, India, is polluted on many levels. Also observed a PLI range of 0.79 to 1.12 in the Atoyac River, Mexico, indicating PTE contamination.

3.3. *Geoaccumulation Index* (I_{geo})

The geoaccumulation index (I_{geo}) was employed to assess sediment metal accumulation, comparing concentrations with undisturbed sediment levels. Calculated using specific formulas, the I_{geo} values indicated pollution levels. Sediment from various river areas exhibited varying levels of contamination, with certain rivers showing moderate Cd contamination and others demonstrating significant Ni contamination.

Examination of the geoaccumulation index (I_{geo}) values revealed that the Buriganga River site was extremely polluted in 2011 due to Pb, Cd, Cr, and Cu contaminants, whereas in 2019 it showed moderate pollution. The Shithalakshya River site was shown to have unpolluted sediment in 2020 while being heavily contaminated by Cd pollution in 2014. The Korotoa River site was moderately polluted in 2022 due to Cd and Pb pollution while it was in an alarming condition in 2015. In contrast, 2015 recorded moderate pollution attributed to Cu, Zn, and Cd contaminants, while 2019 showed an absence of pollution, highlighting the potential ecological health risks (Figure 5). Other rivers, including the Karnaphuli, Khiru, and Dhaleshwari rivers, exhibit moderate levels of Cd contamination. Sediment sourced from the Korotoa River, the Rupsa River, the Bangshi River, and the Buriganga River sites demonstrates moderate Ni contamination. Furthermore, I_{geo} values below 0 are observed in significant river sites, indicating an unpolluted status. Further sediment quality analysis is indispensable to manage and ascertain the origins of pollution for potential ecological health. The River Pra, Ghana showed the moderately to extremely polluted with Pb and Ni. I_{geo} values in the Atoyac River, Mexico indicated unpolluted to moderately polluted with

As, Cu, and Pb. The I_{geo} values in the SomesuMic River, Romania followed as Pb > Cd > Zn > Ni > Cu > Mn > Cr > Fe. Further sediment quality analysis is indispensable to manage and ascertain the origins of pollution for potential ecological health.

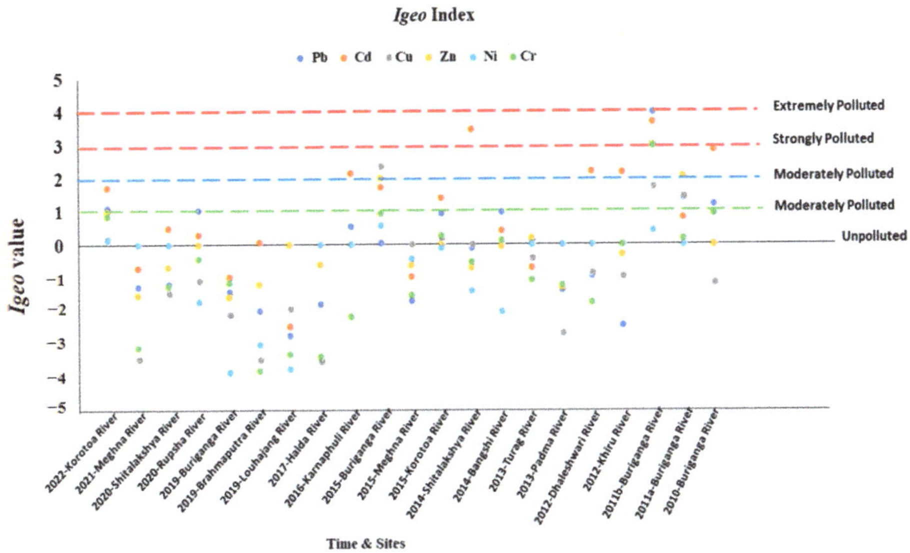

Figure 5. Graph showing geo accumulation index (I_{geo}) value in sediments at different river sites in Bangladesh.

3.4. Potential Ecological Risk (PER)

The incorporation of background values enhances the explanation of geochemical data. The contamination factor (C_f^i) was employed to assess sediment contamination, and the degree of contamination (C_d) was calculated as the sum of all contamination factors. Their calculation utilized specific formulas, incorporating measured heavy metal concentrations and elemental background values. The degree of contamination was categorized into four levels: no pollution, low-to-moderate pollution, significant pollution, and very significant pollution. Pb contamination exhibited temporal variations, with instances of low-to-moderate pollution and substantial pollution in the Buriganga River site (Figure 6). Similar assessments were made for Cd, Cu, Zn, Ni, and Cr, demonstrating varying pollution levels across different river sites. The Cd and Pb concentration found extremely high levels of contamination, which showed very significant pollution $\left(C_f^i \geq 6\right)$ in the sediment of the Shitalakshya and Buriganga rivers.

The potential ecological risk index (PER) evaluated the sensitivity of biological communities to toxic substances. Individual elements, area-wise, underwent risk assessment tests based on specific formulas. There was a considerable risk of Pb contamination in the sediment of the Buriganga River site in 2011, whereas a very high risk (ER $\geq$ 320) was shown for Cd pollution (Figure 6). Except for the Buriganga River, all river sites were revealed as low risk for Pb contamination. A very high risk of Cd contamination existed at the Shitalakshya River site in 2014. The elevated danger of Cd pollution was evident in the rivers Dhaleshwari, Karnaphuli, and Khiru. Additionally, the Bangshi River, Brahmaputra River, Shitalakshya River, and Rupsha River sites were showed considerable risk to moderate risk of Cd pollution. All river sites had a significantly low risk of contamination with Cu, Zn, Ni, and Cr.

River Name	Pollution level (*Cf*) Pb	Cd	Cu	Zn	Ni	Cr	Risk Level (Er) Pb	Cd	Cu	Zn	Ni	Cr	PER
Korotoa River	3.23	4.97	1.69	3.05	1.68	2.76	16.17	149	8.44	3.05	10.07	5.53	192.26
Meghna River	0.62	0.93	0.14	0.53	0	0.18	3.12	28	0.69	0.53	0	0.35	32.69
Shitalakshya River	0.66	2.13	0.55	0.94	0	0.64	3.29	64	2.73	0.94	0	1.28	72.25
Rupsha River	3.12	1.87	0.71	1.52	0.46	1.13	15.6	56	3.55	1.52	2.77	2.26	81.69
Buriganga River	0.57	0.77	0.35	0.51	0.11	0.69	2.85	23	1.77	0.51	0.63	1.38	30.14
Brahmaputra River	0.38	1.6	0.14	0.66	0.19	0.11	1.9	48	0.69	0.66	1.13	0.22	52.6
Louhajang River	0.23	0.28	0.39	0	0.11	0.15	1.15	8.3	1.97	0	0.68	0.31	12.4
Halda River	0.44	0.13	0.13	0.99	0	0.15	2.2	4	0.66	0.99	0	0.29	8.14
Karnaphuli River	2.18	6.7	0	0	0	0.34	10.92	201	0	0	0	0.68	212.6
Buriganga River	1.57	5	7.65	6.02	2.25	2.89	7.85	150	38.24	6.02	13.53	5.78	221.42
Meghna River	0.47	0.77	0	0.99	1.12	0.53	2.37	23	0	0.99	6.72	1.06	34.13
Korotoa River	2.9	4	1.69	0	1.40	1.82	14.5	120	8.44	0	8.38	3.63	154.96
Shitalakshya River	1.42	16.7	0	0.94	0.58	1.05	7.09	501	0	0.94	3.46	2.11	514.6
Bangshi River	3.00	2.03	0	1.46	0.38	1.64	15.00	61	0	1.46	2.27	3.27	83
Turag River	1.64	0.93	1.12	1.74	0	0.72	8.20	28	5.6	1.74	0	1.43	44.97
Padma River	0.59	0	0.24	0.61	0	0.65	2.93	0	1.18	0.61	0	1.30	6.02
Dhaleshwari River	0.79	6.93	0.83	0	0	0.46	3.95	208	4.16	0	0	0.91	217.02
Khiru River	0.28	6.83	0.77	1.22	0	0	1.4	205	3.86	1.22	0	0	211.48
Buriganga River	23.89	19.53	4.99	11.98	2.02	11.82	119.46	586	24.95	11.98	12.12	23.65	778.16
Buriganga River	3.99	2.67	4.10	6.28	0	1.69	19.95	80	20.49	6.28	0	3.37	130.09
Buriganga River	3.49	10.83	0.67	0	2.95	2.91	17.44	325	3.37	0	17.69	5.82	369.31

0	0.9	No pollution (Cif < 1)
1	3	Low to moderate pollution (Cif = 1–3)
3	6	Significant pollution (Cif = 3–6)
6	7	Very significant pollution (Cif ≥ 6)

0	39	Low risk (ER < 40)
40	80	Moderate risk (40 ≤ ER < 80)
80	160	Considerable risk (80 ≤ ER < 160)
160	320	High risk (160 ≤ ER < 320)
320	321	Very high risk (ER ≥ 320)

1	149	low risk (PER <150)
150	299	moderate risk (150 = PER < 300)
300	599	Considerable risk (300 = PER < 600)
600	601	high risk (PER ≥ 600)

Figure 6. Example pollution levels (*Cf*), risk levels (*Er*), and potential ecological risk (PER) levels in river sediment.

We addressed the PER value in the current evaluation of river sediment quality, which is displayed in Figure 6. A moderate-to-extremely high ecological risk was suggested by the PER index level for the sediment from different river sites. Based on the PER value, moderate risk was found in the Korotoa, Karnaphuli, Dhaleshwari, and Khiru rivers, while the Shitalakshya and the Buriganga rivers exhibited significant risk. The PER value showed a moderate risk at the Korotoa River site in 2022. The Buriganga River (PER $\geq$ 600) and the Shitalakshya River (300 $\leq$ PER < 600) sites showed higher potential ecological risk than other rivers. Both rivers are located in the industrial area of Dhaka. Pb and Cd, especially, enhanced the potential ecological risk. At these two river locations, different sources contribute to the PTE pollution of the sediment. The PER values in the Atoyac River, Mexico showed decreasing order: As (80) > Pb (70) > Cu (57) > Ni (38) > Cr (26) > Zn (11).

4. PTE Pollution Status in River Water

The majority of urban river water is contaminated by debris from anthropogenic activity, chemical discharge from industrial zones, and wastewater. In Bangladesh, the Louhajang River water showed safe level of PTE contamination [92]. Many researchers have significantly evaluated PTE concentrations in different river water [93–99] (Table 4). During the winter the mean concentration of PTEs in Buriganga river water followed Fe > Cr > Ni > Zn > Cu > Pb, but in the summer, the order was Fe > Cr > Zn > Ni > Cu > Pb. The highest level of Fe in the Buriganga River was recorded to be 20.43 mg/L in wintertime but dropped to 13.49 mg/L in summertime. The readings in the Turag River were 10.16 mg/L and 7.1 mg/L, respectively. Among the rivers on the list, Buriganga, Turag, Dhaleshwari, Korotoa, Karnaphuli, and Bangshi all had Cr contamination in their river water. High concentrations of Ni were discovered in water from the Lee River, England and the Ganga River, India (Table 4).The Turag river water showed high level of PTE contamination [93], whereas Shitalakhya River water showed the safe level of PTE pollution [94].

Table 4. Pollution source and PTE concentrations in river waters from different areas.

Time	Water Body Name	River Water Study Code	Country/ City Code	Pollution Source	Cr	Mn	Fe	Ni	Cu	Zn	As	Pb	Cd	Unit
2021	Buriganga River	E1	B/D	TDIMW	1.99	-	12.31	1.05	0.685	1.065	-	0.27	-	mg/L
2021	Turag River	E2	B/D	TDIMW	0.61	-	6.995	0.965	0.75	1.6	-	0.32	-	mg/L
2021	Halda River	E3	B/C	IESG	0.032	0.08	0.48	-	1.751	0.2275	1.205	0.067	0.036	mg/L
2021	Karnaphuli River	E4	B/C	WBWMS	0.006	1.53	4.34	0.018	0.007	0.059	-	0.022	0.017	mg/L
2021	Louhajang River	E9	B/D	IMHA	0.0052	-	-	0.0041	0.0062	-	0.0059	0.0051	0.0004	mg/L
2021	Dhaleshwari River	E10	B/D	IMDW	0.71	-	-	0.62	-	0.18	-	-	0.19	mg/L
2021	Meghna River	E11	B/C	IMDW	0.045	-	-	-		-	0.024	0.009	0.018	mg/L
2021	Sela River	E12	B/K	OSAF	0.0346	0.035	0.0734	0.1481	0.0236	0.028	-	0.0837	0.0307	mg/L
2020	Ganges River	E13	B/K	Geogenic	0.0007	0.0002	0.08	0.005	0.0067	0.089	0.0021	0.0026	0.0001	mg/L
2019	Old Brahmaputra	E6	B/D	IDID	0.01	1.44	-	0.44	0.12	0.01	-	0.11	0.001	mg/L
2019	Ganges River	E8	B/R	IMA	0.038			0.004	0.012	0.03	0.003	0.009		mg/L
2018	Korotoa River	E14	B/R	MIEAR	1.13	-	-	1.33	3.02		2.62	0.81	0.75	mg/L
2018	Balu River	E15	B/D	IMDW	0.0044			0.0132			-	0.0186	0.003	mg/L
2017	Passur River	E16	B/K	GS	0.02		0.27		0.02	0.01				mg/L
2016	Turag River	E17	B/D	IMDW	0.0558	0.7085	2.606	0.1309	0.2308	0.2952	0.0034	0.0146	0.0143	mg/L
2016	Shitalakhya River	E18	B/D	IMDW					0.0242			0.023	0.0073	mg/L
2016	Karnaphuli River	E5	B/C	WBWMS	69.56						23.36	9.85	6.46	µg/L
2015	Buriganga River	E19	B/D	TDIMW	0.114	0.157	0.612	0.15	0.239	0.332	0.134	0.119	0.059	mg/L
2015	Dakatia River	E20	B/C	USF	0.003	0.334	0.218		0.033	0.114		0.006	0.001	mg/L
2013	Bangshi River	E21	B/D	TDAD	0.093	0.088		0.035	1.05	3.32	0.024	0.108	0.007	mg/L
2012	Khiru River	E7	B/M	IMDW		0.167			0.0043	0.007		0.0221	0.128	mg/L
	Ganga River	W10	India/		70.16			61.11	43.72	71.37		80.55	11.41	µg/L
	Lee River	W5	England/					55.1	46.7	62.7		13.3	0.44	µg/L
	Yangtze River	W15	China/		1.3				2.8	31		2	0.4	µg/L
	BDSW	E22			0.05	0.1	0.3	0.1	1	5	0.05	0.05	0.005	mg/L
	WHO	E23			5						10	10	3	µg/L
	TRV				11						150	3	2	µg/L

Note: B-Bangladesh, D: Dhaka, C: Chittagong, K: Khulna, M: Mymensingh, R: Rajshahi. TDIMW: textile and dyeing industries, municipal waste; IDID: industrial, domestic, and irrigation discharges; IMDW: industrial and municipal discharged water; IMHA: industrial, municipal, and household activities; TDAD: textile, dyeing, and apparel industries, DEPZ; USF: urban sewage, different types of factories; IESG: industrial effluents, sewage, geogenic reason; WBWMS: wastewater, bedrock weathering, metal smelting; GS: geological sources; OSAF: oil spillage, accident, factories; MIEAR: municipal, industrial effluents, agricultural runoff; BDSW: Bangladesh's standard of water quality. Here, E1 [22], E2 [22], E3 [23], E4 [24], W10 [54], E5 [59], E6 [70], E7 [80], E8 [90], W5 [87], W15 [91], E9 [92], E10 [95], E11 [96], E12 [97], E13 [98], E14 [99], E15 [100], E16 [101], E17 [93], E18 [94], E19 [102], E20 [103], E21 [104], E22 [105], E23 [106].

The water of Dakatia, Khiru, Old Brahmaputra, and Karnaphuli all had signs of Mn poisoning. The levels of Fe contamination in the Karnaphuli, Dhaleshwari, Turag, and Buriganga waterways rose. The Buriganga and Korotoa Rivers exhibited high levels of Ni contamination, above Bangladesh's water quality standard guideline values. Water from the Bangshi, Korotoa, and Halda rivers had high levels of Cu contamination. The Karnaphuli, Korotoa, and Halda rivers were found to be contaminated by As. Water from the Karnaphuli River exhibited high levels of Pb and Cd contamination. Similarly, high levels of Pb and Cu pollution in water were shown in the Ganga River in India, the Lee River in England, and the Yangtze River in China (Table 4). The Buriganga [102] and Dakatia [103] river water showed higher Mn content than the normal content that reported by BDSW. The water of the Bangshi River had a Cr concentration of 0.093 mg/L [104], whereas BDSW [105] recorded 0.05 mg/L.WHO [106] found that the content of Cr in water is 5 μg/L.

5. PTE Pollution Status in Aquatic Species

Numerous researches have discovered PTE contamination in fish, crabs, and molluscs, and small amounts of PTEs may build up in the river environment in fish and other aquatic species. High levels of Cr were found in fish and shellfish of the river Buriganga and Turag (Table 5). Safe level of Pb and Cd content was reported in fish from Kelantan River, Malaysia [107]. Khan et al. [108] reported high levels of Pb, Cd, and Cr in freshwater fish at the Swat River, Pakistan. Scientifically, many researcher reported notable level of PTE concentration in aquatic species [109–119] from the river ecosystem. PTE pollution in the sediment affects the bottom-feeder aquatic species (*A. coila, G. youssoufi, M. pancalus, H. fossilis, C. punctata, M. vittatus*). Also Concentration of PTE reported in crustaceans (*M. rosenbergii*), and molluscs (*I. exustus*) (Table 5). Human ingestion of these type of species from the polluted river habitats increases the risk of cancer. Health hazards have been connected to PTE contamination of fish and shellfish from sediment areas. Cu exposure from sediment caused physiological stress in organisms, which impacted their food chain, general well-being, and cellular functions. Diseases and other environmental stresses are more likely to affect organisms due to Cu pollution. Fish species from the Dhaleswari, Buriganga, and Turag rivers showed high levels of Cr contamination. Fish species in the Turag River displayed high levels of Fe contamination. All of the mentioned rivers' fish, crustaceans, and mollusc species had high levels of Zn contamination, except the Turag-Tongi-Balu River. The fish species found in the Karnaphuli and Bangshi rivers had high levels of As pollution.The majority of the fish and mollusc species in the river exhibited significant levels of Pb pollution. Fish and crab species from the Buriganga, Turag-Tongi-Balu, and Bangshi rivers exhibited high levels of Cd contamination.

Table 5. PTE concentration (mg/kg) in fish, crustaceans, and molluscs from different rivers.

Time	Name	Country (City Code)	Study Code	Aquatic Species	Analytical Method	Cr	Mn	Fe	Ni	Cu	Zn	As	Pb	Cd
2024	Turag-Tongi-Balu	Bangladesh (D)	S1	*H. fossilis*	ICP-MS and FAAS	1.74	26.32	19	1.34	0.2	6.39	0.16	4.17	0.67
2023	Swat River	Pakistan	S2	fish	ICP-MS	14.59			11.80	8.59	116.83		1.95	0.29
2022	Dhaleswari River	Bangladesh(D)	S3	*H. fossilis*	AAS	22.6			5.55		98.19		7	
2021	Buriganga River	Bangladesh (D)	S4	*H. fossilis*	AAS	187.07		39.07	3.01	3.51	35.12		5.07	
2021	Turag River	Bangladesh (D)	S5	*H. fossilis*	AAS	70.18		45.1	12.18	6.03	68.25		6.22	
2019	Karnaphuli River	Bangladesh(C)	S6	*P. chinensis*	ICP-MS	3.59				13.1		5.03	14	0.44
2019	Karnaphuli River	Bangladesh (C)	S7	*T. ilisha*	GF-AAS and AAS	0.65						1.22	0.67	0.15
2016	Buriganga River	Bangladesh(D)	S8	*L. rohita*	ICP-MS	18.84	125.81		6.64	18.77	251.69	0.73	6.98	0.04
2015	Buriganga River	Bangladesh(D)	S9	*M. pancalus*	ICP-MS	7.18	25.65		1.6	11.66	165.1	0.22	3.17	0.01
2015	Buriganga River	Bangladesh(D)	S9	*M. rosenbergii* *	ICP-MS	1.59	35.25		0.44	575.34	187.04	1.19	0.51	1.51
2015	Buriganga River	Bangladesh(D)	S9	*I. exustus* **	ICP-MS	16.05	319.66		5.75	16.47	58.56	1.02	4.55	0.05
2014	Kelantan River	Malaysia	S12	fish	GF-AAS								0.072	0.05
2012	Bangshi River	Bangladesh(D)	S13	*H. fossilis*	AAS	0.71	26.11		4.11	14.17	203.19	6.24	7.71	0.31
	USFDA (FSG)		S14			13			80					
	FAO		S15			1	1			10	30	1	2.5	0.2
	WHO		S16							30			2	
	FAO/WHO		S17									1		
	IOM		S18					40–45						
	EC (EU)		S19			1				3			0.1	0.05

Note: D: Dhaka, C: Chittagong, * crustaceans, ** molluscs. Here, S7 [6], S4 [22], S5 [22], S1 [35], S12 [107], S2 [108], S3 [109], S6 [110], S8 [111], S9 [112], S13 [113], S14 [114], S15 [115], S16 [116], S17 [117], S18 [118], S19 [119].

6. Effects on Human Health

It is not possible for the public to completely avoid the potential health concerns associated with consuming aquatic animals that have accumulated PTEs. These PTEs disturb the food pyramid and endanger human health by creating cancer and human organ disorders.Hazard index (HI) value of freshwater fish greater than 1 (HI > 1) indicates non carcinogenic risk to human health (adult). China's freshwater fish and crabs' hazard index (HI) value is in the safe region (<1) in Table 6. High values of hazard index (HI >1) reported in *L. rohita* (3.78) (Bangladesh), *O. niloticus* (4.53) (Egypt), *P. notialis* (6.1) (Ghana), *M. armatus* (13.71) (India), and *P. laevis* (1.5) (China) from the river ecosystems that related to PTE pollution.

Many researchers have reported cancer risk (CR) values of Ni, As, Cd, and Pb, as shown in Table 7 for different aquatic species. CR values ranging from 10^{-5}–10^{-4} for all metals are related to moderate cancer risk, and 10^{-3}–10^{-1} indicates cancer risk in human health. Neurological diseases in humans are caused by high amounts of Cd, Pb, and Cr [120,121]. Eye yellowing is a result of As, Cd, and Pb pollution and decreased thyroid hormone synthesis [122], elevated heart illnesses [123], and exacerbated peripheral vascular disease [124]. Contamination with Cr, As, Cd, and Pb leads to haemolysis in illnesses of the cardiovascular system [125,126]. Cd pollution causes illnesses such as pulmonary system fibrosis [127] and bone degradation [128]. Long-term exposure to Cd has been linked to harmful health outcomes, such as lung and kidney damage, as well as an increased chance of developing several malignancies.

According to Kim et al. [129], liver disorders and cirrhosis are caused by Pb, As, and Cd pollution. Contamination with Cd, Pb, As, and Cr can cause skin illness and gastrointestinal upset [130]. Tiredness, feeling sick, haemophilia, and electrolyte imbalance have all been linked to Zn toxicity. Long-term exposure to inorganic As can have negative consequences on the neurological system, haematological system, skin, liver, gastrointestinal tract, respiratory tract, and cardiovascular system [131].Women will have difficulties becoming pregnant [132,133] due to Pb, Cd, and As contamination. For human health, the NYSDOH [134] categorises cancer risk (CR) as follows: if CR $\leq 10^{-6}$ = low; 10^{-4} to 10^{-3} = moderate; 10^{-3} to 10^{-1} = high; $\geq 10^{-1}$ = very high. We have found different toxicity effects on human health due to PTE pollution in wetland ecosystems (Figure 7).

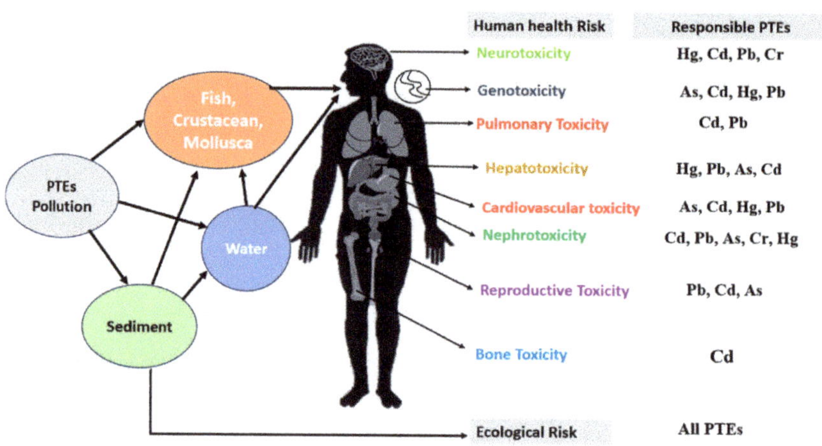

Figure 7. A model of PTE pollution in river sediment, water, and aquatic species, and showing possible toxicity risk in human health for different PTEs.

Table 6. Hazard index (HI) values reported for aquatic species.

Species	Study Code	Hazard Index (HI)	Country
L. rohita	H1	3.78	Bangladesh
C. punctata	H1	1.72	Bangladesh
C. batrachus	H1	1.17	Bangladesh
H. fossilis	H1	1.10	Bangladesh
Cynoglossus sp.	H5	23.57	Nigeria
Fish sp.	H6	1.69	China
C. fluminea	H7	1.67	Bangladesh
C. amnicola	H8	8.34	Nigeria
Clam	H9	1.148	China
Freshwater fish	H10	0.558	China
Freshwater crab	H10	0.092	China

Note: here, H1 [35], H5 [135], H6 [136], H7 [137], H8 [138], H9 [139], H10 [140].

Table 7. CR value of PTEs from consumption of aquatic species.

Species	Study Code	Potential Toxic Elements (PTEs)				Country
		As	Pb	Ni	Cd	
L. rohita	A1	2.86×10^{-5}	1.76×10^{-6}	3.00×10^{-4}	8.32×10^{-5}	Bangladesh
C. punctata	A1	3.08×10^{-5}	5.69×10^{-6}	9.73×10^{-5}	3.00×10^{-3}	Bangladesh
H. fossilis	A1	3.52×10^{-5}	5.20×10^{-6}	3.00×10^{-4}	6.00×10^{-4}	Bangladesh
Cynoglossus sp.	B1		4.02×10^{-8}	2.35×10^{-5}	6.48×10^{-6}	Nigeria
C. fluminea	C1	1.00×10^{-4}	1.00×10^{-6}	2.00×10^{-4}	2.00×10^{-3}	Bangladesh
C. amnicola	F1	1.09×10^{-7}	1.66×10^{-8}		4.64×10^{-8}	Nigeria
L. camtschaticum	G1	1.60×10^{-4}				China
C. farreri	H1		7.44×10^{-8}		1.93×10^{-3}	China
C. ariakensis	H1		5.87×10^{-8}		2.38×10^{-4}	China
S. constricta	H1		7.26×10^{-8}		3.94×10^{-5}	China
C. gariepinus	K1	2.00×10^{-3}	9.90×10^{-6}			Egypt
O. niloticus	K1	1.40×10^{-3}	1.20×10^{-5}			Egypt
O. aureus	K1	1.50×10^{-3}	9.80×10^{-6}			Egypt
T. zillii	K1	1.70×10^{-3}	9.60×10^{-6}			Egypt
H. molitrix	P1		7.11×10^{-6}	2.55×10^{-5}		China
C. idellus	P1		1.97×10^{-6}	1.85×10^{-4}		China

Note: here, A1 [35], B1 [135], C1 [137], F1 [138], G1 [141], H1 [142], K1 [143], P1 [144].

7. Research Perspective: Implications and Future Directions

The assessment of pollution risk through river sediments, water, and aquatic species is highly important for ecosystems. This review found different river sites have a spectrum of pollution levels, ranging from moderate to strong ecological risk. Notably, the Buriganga River and Shitalakshya River sites stand out as areas of heightened ecological risk, surpassing other locations. This disparity underscores the urgency of in-depth investigations and continuous monitoring in these critical zones, necessitating further research to comprehend their intricate dynamics and the potential ramifications for river ecology. Furthermore, the examination of geoaccumulation index values across diverse river sites has consistently revealed levels exceeding background sediment values. Identified potential ecological risks (PERs) at the Buriganga River and Shitalakshya River sites flag them as high-risk zones, raising concerns about their impact on benthic organisms, aquatic species (fish, crustaceans, molluscs), and even human health. These findings underscore the interconnections of ecological well-being and human health safety, highlighting the imperative of comprehensive ecosystem management. The observed spatial disparities in PTE pollution and sediment accumulation underline the intricate environmental dynamics within river ecosystems. The breach of LELs for specific metals in certain regions signals a potential menace to the ecological equilibrium of these areas. Cd and Pb concentrations exceeding typical shale values raise alarm bells, demanding immediate attention to mitigate potential adverse repercussions on the environment and human health.

This review offers crucial insights into the ecological risks posed by PTE accumulation in river sediments, water, and aquatic species. The divergent pollution levels across various river sites and the intensified health risk in specific regions underscore the necessity for ongoing research on sediment, water, and aquatic species; proactive surveillance; and strategic interventions to ensure river ecosystems' long-term vitality and sustainability. The findings highlight the pivotal role of interdisciplinary collaboration and well-informed decision making in mitigating the potential impact of PTE contamination on the environment and human populations.

Potential toxic elements (PTEs) are released into the riverbank ecosystem and pollute river sediment, water, and aquatic species as a consequence of several industrial processes, including municipal trash, fuel refining, smelting, tannery waste, and chemical waste. An ecological risk assessment of sediment encompasses evaluation of potential adverse effects stemming from contaminants or stressors on sediment ecosystems and their inhabitants. Sediments serve as repositories for diverse pollutants, including PTEs, organic chemicals, and nutrients. Accumulation of these contaminants over time can endanger aquatic life, benthic organisms, and even humans through the food chain. This review details the extent of sediment, water, and aquatic species pollution in the river area. The presence of sediment pollution amplifies ecological risks, with approximately 83% of water bodies exhibiting high pollution rates. Across all the studied rivers, the average concentration of various PTEs (Pb, Cd, Cr, Cu, Zn, Ni) in sediment exceeded recommended Sediment Quality Guidelines (SQGs) ranges, following the order: Zn > Cr > Ni > Cu > Pb > Cd. Pollution Load Index (PLI > 1), PER index, and geoaccumulation (I_{geo}) index values collectively designate the Buriganga, Turag, Korotoa, Karnaphuli, Rupsha, and Shitalakshya river sites as heavily polluted due to PTE contamination. Effective management of pollutants is of paramount importance for minimizing the ecological impact of hazardous industrial materials and contaminants. Consequently, this study identifies, discusses, and underscores potential ecological risks using various risk assessment methodologies and established risk thresholds. There are concerns about human health due to PTEs' contamination of water and aquatic species. Monitoring the river region's environment necessitates continued research on sediment, water quality, and pollution dynamics, serving as a valuable foundation for future studies in this domain.

8. Conclusions

Anthropological activities and industrial discharges are causes of the alarming levels of PTE pollution in urban river ecosystems. Highly industrialised sites and residential settlements next to release sources raise exposure concerns. Our findings showed that risk models significantly underestimate the quantities of PTEs in river sediment. Concerning levels of PTEs were found in the sediment, water, and aquatic species, which has a significant impact on the environment and human health. Human health toxicity risks were emphasised by PTEs, especially As, Pb, and Cd contamination. Also indicated carcinogenic risk through CR value. An hazard index value greater than 1 poses risk to human health. High content of Mn poses non-carcinogenic risk in human health. A number of rivers showed ecological risk associated with PTEs. These findings highlight the value of innovation as a tool for regulatory monitoring for precisely identifying hazardous exposures, pollution levels, and ecological risk levels in river sediments. The PLI, I_{geo}, and PER revealed that river sediments exhibit significant Cd and Pb contamination, particularly in the Buriganga River, posing serious consequences and risks to human health and aquatic life. The PER for the Shitalakshya and Buriganga rivers is very high, indicating severe ecological risks in addition to medium-level contamination of the Dhaleshwari and Khiru rivers. The environmental risk was elevated by PTE pollution in the riverbed region, which demonstrated moderate-to-significant ecological risks as well as contamination of water and aquatic species. Specifically, the additive effects exacerbated by the presence of Mn, Zn, Pb, Cd, and Ni in aquatic species; Fe, Ni, and As in water; and Pb, Cd, Cr, and Ni in sediment further aggravate ecological risks in aquatic ecosystems, emphasizing the crucial need for continuous monitoring and stricter regulations to safeguard the sustainability of river ecosystems and protect public health we can make significant progress in mitigating the ecological and human health risks by continuous monitoring, combined with proactive management and policy interventions, water treatment, sustainability and health of these vital water resources. In addition, strengthening the environmental regulations to enforce stricter limits on PTE emissions from industrial and urban sources and the implementation of reformed policies could incentivise the adoption of cleaner technologies and penalise non-compliance. Finally, educating the masses about the ecological consequences of PTE contamination and encouraging industries to adopt sustainable practices, such as closed-loop systems and waste minimization techniques, could significantly reduce the discharge of PTEs into the environment.

Author Contributions: M.M.H.—conceptualization, methodology, paper screening, extraction, data curation and formal analysis, visualization, primary and original draft writing, editing; I.J.—review and resources; M.A.D. and M.J.D.—review and editing; S.J.M.—review, editing and resources; A.F.B.M.—resources, editing; S.H.—review, editing; D.Z.—supervision, conceptualization, editing and review. All authors have read and agreed to the published version of the manuscript.

Funding: The Key Research and Development Program of Jiangsu Province (BE2021691), and the National Key R & D Program of China (2023YFC3403600).

Data Availability Statement: Data will be made available on request.

Acknowledgments: Thanks to Md Abdul Quadir and Shakhina Khatun for their valuable supports, Also thanks to Nyma Sultana and Mariam Hossain for valuable support in completing the study.

Conflicts of Interest: All authors declare that there is no conflicts of interest.

References

1. Khalijian, A.; Lorestani, B.; Sobhanardakani, S.; Cheraghi, M.; L Tayebi, L. Ecotoxicological Assessment of Potentially Toxic Elements (as, Cd, Ni and V) Contamination in the Sediments of Southern Part of Caspian Sea, the Case of Khazar Abad, Mazandaran Province, Iran. *Bull. Environ. Contam. Toxicol.* **2022**, *109*, 1142–1149. [CrossRef]
2. Liu, X.; Dadzie, A.A.; Yuan, L.; Xing, S.; Zhou, X.; Xiao, S. Analysis and potential ecological risk assessment of heavy metals in surface sediments of the freshwater ecosystem in Zhenjiang City, China. *SN Appl. Sci.* **2022**, *4*, 258. [CrossRef]
3. Hossain, M.B.; Sultana, J.; Pingki, F.H.; Nur, A.A.U.; Mia, M.S.; Bakar, M.A.; Yu, J.; Paray, B.A.; Arai, T. Accumulation and contamination assessment of heavy metals in sediments of commercial aquaculture farms from a coastal area along the northern Bay of Bengal. *Front. Environ. Sci.* **2023**, *11*, 1148360. [CrossRef]
4. Bangladesh Economic Review. Ministry of Finance of Government of Bangladesh. 2023. Available online: http://mof.gov.bd (accessed on 20 June 2023).
5. Mahamat, T.N.; Kidjakebo, S.; Abel, S.K.; Bertrand, N.G. Heavy metals pollution in water, sediments and fish of Léré Lake, Western Mayo Kebbi (Chad). *Int. J. Environ. Prot. Pollut.* **2017**, *5*, 1–11. [CrossRef]
6. Ali, H.; Khan, E.; Ilahi, I. Environmental Chemistry and ecotoxicology of hazardous heavy metals: Environmental persistence, toxicity, and bioaccumulation. *Hindawi. J. Chem.* **2019**, *14*, 6730305. [CrossRef]
7. Reza, R.; Singh, G. Heavy metal contamination and its indexing approach for river water. *Int. J. Environ. Sci. Technol.* **2010**, *7*, 785–792. [CrossRef]
8. Bain, D.J.; Yesilonis, I.D.; Pouyat, R.V. Metal concentrations in urban riparian sediments along an urbanization gradient. *Biogeochemistry* **2012**, *107*, 67–79. [CrossRef]
9. Wang, Z.; Yao, L.; Liu, G.; Liu, W. Heavy metals in water, sediments and submerged macrophytes in ponds around the Dianchi Lake, China. *Ecotox. Environ. Safe* **2014**, *107*, 200–206. [CrossRef] [PubMed]

10. Diagomanolin, V.; Farhanga, M.; Ghazi-Khansaria, M.; Jafarzadehb, N. Heavy metals (Ni, Cr, Cu) in the Karoon waterway river, Iran. *Toxicol. Lett.* **2004**, *151*, 63–67. [CrossRef]
11. Ramessur, R.T.; Parry, S.J.; Ramjeawon, T. The relationship of dissolved Pb to some dissolved trace metals (Al, Cr, Mn and Zn) and to dissolved nitrate and phosphate in a freshwater aquatic system in Mauritius. *Environ. Int.* **2001**, *26*, 223–230. [CrossRef] [PubMed]
12. Shakil, M.S.Z.; Mostafa, M.G. Water quality assessment of paper mills effluent discharge areas. *Al Nahrain J. Sci. ANJS* **2021**, *24*, 63–72. [CrossRef]
13. Shakil, M.S.Z.; Mostafa, M.G. Paper industries concern water pollution: A review. *Int. J. Innov. Res. Rev.* **2021**, *9*, 19–31.
14. Rahim, M.A.; Mostafa, M.G. Impact of Sugar Mills Effluent on Environment around Mills Area. *AIMS Environ. Sci.* **2021**, *8*, 86–99. [CrossRef]
15. Theofanis, Z.U.; Astrid, S.; Lidia, G.; Calmano, W.G. Contaminants in sediments: Remobilisation and demobilization. *Sci. Total Environ.* **2001**, *266*, 195–202.
16. Monira, U.; Sattar, G.S.; Mostafa, M.G. Characterization of tannery effluent of savar tannery estate in Bangladesh. *BAUET J.* **2022**, *3*, 67–76.
17. Islam, M.S.; Mostafa, M.G. Influence of chemical fertilizers on arsenic mobilization in the alluvial Bengal delta plain: A critical review. *J. Water Supply Res. Technol. Aqua* **2021**, *70*, 948. [CrossRef]
18. Monira, U.; Sattar, G.S.; Mostafa, M.G. Characterization and removal efficiency of central effluent treatment plant (CETP). *J. Sustain. Environ. Manag.* **2023**, *2*, 42–50. [CrossRef]
19. Chowdhury, M.; Mostafa, M.G.; Biswas, T.K.; Saha, A.K. Treatment of leather industrial effluents by filtration and coagulation processes. *J. Water Resour. Ind.* **2013**, *3*, 11–22. [CrossRef]
20. Kubra, K.; Mondol, A.H.; Ali, M.M.; Palash, M.A.U.; Islam, M.S.; Ahmed, A.S.S.; Masuda, M.A.; Islam, A.R.M.T.; Bhuyan, M.S.; Rahman, M.Z.; et al. Pollution level of trace metals (As, Pb, Cr and Cd) in the sediment of Rupsha River Bangladesh: Assessment of ecological and human health risks. *Front. Environ. Sci.* **2022**, *10*, 778544. [CrossRef]
21. Bashar, T.; Fung, I.W.H. Water Pollution in a Densely Populated Megapolis, Dhaka. *Water* **2020**, *12*, 2124. [CrossRef]
22. Hossain, M.N.; Rahaman, A.; Hasan, M.J.; Uddin, M.M.; Khatun, N.; Shamsuddin, S.M. Comparative seasonal assessment of pollution and health risks associated with heavy metals in water, sediment and Fish of Buriganga and Turag River in Dhaka City, Bangladesh. Springer. *SN Appl. Sci.* **2021**, *3*, 509. [CrossRef]
23. Rakib, M.R.J.; Jolly, Y.N.; Begum, B.A.; Choudhury, T.R.; Fatema, K.J.; Islam, M.S. Assessment of trace element toxicity in surface water of a fish breeding river in Bangladesh: A novel approach for ecological and health risk evaluation. *Toxin Rev.* **2021**, *41*, 420–436. [CrossRef]
24. Hasan, M.F.; Nur-E-Alam, M.; Salam, M.A.; Rahman, H.; Paul, S.C.; Rak, A.E.; Ambade, B.; Towfiqul Islam, A.R.M. Health risk and water quality assessment of surface water in an urban river of Bangladesh. *Sustainability* **2021**, *13*, 6832. [CrossRef]
25. BBS. *Survey of Manufacturing Industries (SMI) 2012*, 1st ed.; Bangladesh Bureau of Statistics (BBS): Dhaka, Bangladesh, 2013.
26. Proshad, R.; Kormoker, T.; Mursheed, N.; Islam, M.M.; Bhuyan, M.I.; Islam, M.S. Heavy metal toxicity in agricultural soil due to rapid industrialization in Bangladesh: A review. *Int. J. Adv. Geosci.* **2018**, *6*, 83–88. [CrossRef]
27. Sakib, S.M.N. Garment Industry in Bangladesh Struggles to Contain Pollution. The Third Pole. 2023. Available online: https://www.thethirdpole.net/en/pollution/garment-industry-bangladesh-struggles-to-contain-pollution/ (accessed on 12 January 2023).
28. Kamruzzaman, M.; Sakib, S.M.N. Chemicals, industrial waste contamination turn 6 Bangladesh rivers untreatable, Anadolu Ajansı. Available online: https://www.aa.com.tr/en/asia-pacific/chemicals-industrial-waste-contamination-turn-6-bangladesh-rivers-untreatable/2534527#:~:text=According%20to%20official%20records%20of,greater%20Dhaka%20and%20adjacent%20areas (accessed on 14 March 2022).
29. Ruba, U.B.; Chakma, K.; Senthi, J.; Rahman, S. Impact of Industrial Waste on Natural Resources: A Review in the Context of Bangladesh. *Curr. World Environ.* **2021**, *16*, 348–361. [CrossRef]
30. DoE. Department of Environment. *Industrial Waste Management in Bangladesh: An Evaluation on Policy Implementation*; Annual Report; Ministry of Environment and Forest: Dhaka, Bangladesh, 2016.
31. Arikibe, J.E.; Prasad, S. Determination and comparison of selected heavy metal concentrations in seawater and sediment samples in the coastal area of Suva, Fiji. *Mar. Pollut. Bull.* **2020**, *157*, 111157. [CrossRef] [PubMed]
32. USEPA. *Methods for Collection, Storage and Manipulation of Sediments for Chemical and Toxicological Analyses: Technical Manual*; EPA-823-B-01-002; Office of Water: Washington, DC, USA, 2001.
33. Nargis, A.; Sultana, S.; Raihan, M.J.; Haque, M.; Sadique, A.; Sarkar, M.K.; Un-Nabie, M.M.; Zhai, W.; Cai, M.; Habib, A. Multielement analysis in sediments of the River Buriganga (Bangladesh): Potential ecological risk assessment. *Int. J. Environ. Sci. Technol.* **2019**, *16*, 1663–1676. [CrossRef]

34. Kamzati, L.L.J.; Kaonga, C.C.; Mapoma, H.W.T.; Thulu, F.G.; Abdel-dayem, S.M.; Anifowose, A.J.; Chidya, R.C.G.; Chitete-Mawenda, U.; Sakugawa, H. Heavy metals in water, sediment, fish and associated risks from an endorheic lake located in Southern Africa. *Int. J. Environ. Sci. Technol.* **2020**, *17*, 253–266. [CrossRef]
35. Hossain, M.M.; Jahan, I.; Al Nahian, A.; Zhuang, Z.; Maxwell, J.S.; Ali, M.Y.; Sethupathy, S.; Zhu, D. Immediate health risk: Concentration of heavy metals in contaminated freshwater fishes from the river channel of Turag-Tongi-Balu. *Environ. Toxicol.* **2024**, *39*, 120–134. [CrossRef]
36. Taylor, S.R. Abundance of chemical elements in the continental crust: A new table. *Geochim. Cosmochim. Acta* **1964**, *28*, 1273–1285. [CrossRef]
37. Hilton, J.; Davison, W.; Ochsenbein, U. A mathematical model for analysis of sediment coke data. *Geology* **1985**, *48*, 281–291.
38. Karadede, H.; Unlu, E. Concentrations of some heavy metals in water, sediment and fish species from the Atatürk Dam Lake (Euphrates), Turkey. *Chemosphere* **2000**, *41*, 1371–1376. [CrossRef]
39. Suresh, G.; Ramasamy, V.; Meenakshisundaram, V.; Venkatachalapathy, R.; Ponnusamy, V. Influence of mineralogical and heavy metal composition on natural radionuclide concentrations in the river sediments. *Appl. Radiat. Isot.* **2011**, *69*, 1466–1474. [CrossRef] [PubMed]
40. Müller, G. Index of geoaccumulation in sediments of the Rhine River. *Geo J.* **1969**, *2*, 108–118.
41. Rabee, A.M.; Al-Fatlawy, Y.F.; Nameer, M. Using pollution load index (PLI) and geo accumulation index (I_{geo}) for the assessment of heavy metals pollution in Tigris river sediment in Baghdad Region. *J. Al Nahrain Univ.* **2011**, *14*, 108–114. [CrossRef]
42. Hakanson, L. An ecological risk index for aquatic pollution control a sedimentological approach. *Water Res.* **1980**, *14*, 975–1001. [CrossRef]
43. Guo, W.; Liu, X.; Liu, Z.; Li, G. Pollution and potential ecological risk evaluation of heavy metals in the sediments around Dongjiang Harbor, Tianjin. *Procedia Environ. Sci.* **2010**, *2*, 729–736. [CrossRef]
44. Liu, C.; Xu, J.; Liu, C.; Zhang, P.; Dai, M. Heavy metals in the surface sediments in Lanzhou Reach of Yellow River, China. *Bull. Environ. Contam. Toxicol.* **2009**, *82*, 26–30. [CrossRef] [PubMed]
45. Huang, Z.; Liu, C.; Zhao, X.; Dong, J.; Zheng, B. Risk assessment of heavy metals in the surface sediment at the drinking water source of the Xiangjiang River in South China. *Environ. Sci. Eur.* **2020**, *32*, 23. [CrossRef]
46. Chai, L.; Li, H.; Yang, Z.; Min, X.; Liao, Q.; Liu, Y.; Men, S.; Yan, Y.; Xu, J. Heavy metals and metalloids in the surface sediments of the Xiangjiang River, Hunan, China: Distribution, contamination, and ecological risk assessment. *Environ. Sci. Pollut. Res.* **2017**, *24*, 874–885. [CrossRef] [PubMed]
47. Singh, K.P.; Malik, A.; Sinha, S.; Singh, V.K.; Murthy, R.C. Estimation of source of heavy metal contamination in sediments of Gomti river (India) using principal component analysis. *Water Air Soil Pollut.* **2005**, *166*, 321–341. [CrossRef]
48. Alyazichi, Y.M.; Jones, B.G.; Mclean, E.; Pease, J.; Brown, H. Geochemical assessment of trace element pollution in surface sediments from the Georges River, Southern Sydney, Australia. *Arch. Environ. Contam. Toxicol.* **2016**, *72*, 247–259. [CrossRef]
49. Filgueiras, A.V.; Lavilla, I.; Bendicho, C. Evaluation of distribution, mobility and binding behaviour of heavy metals in surficial sediments of Louro River (Galicia, Spain) using chemometric analysis: A case study. *Sci. Total. Environ.* **2004**, *330*, 115–129. [CrossRef] [PubMed]
50. Kuriata-Potasznik, A.; Szymczyk, S.; Skwierawski, A.; Glinska-Lewczuk, K.; Cymes, I. Heavy metal contamination in the surface layer of bottom sediments in a flowthrough lake: A case study of Lake Symsar in Northern Poland. *Water* **2016**, *8*, 358. [CrossRef]
51. Brügmann, L. Metals in sediments and suspended matter of the river Elbe. *Sci. Total Environ.* **1995**, *159*, 53–65. [CrossRef]
52. Proshad, R.; Kormoker, T.; Abdullah Al, M.; Islam, M.S.; Khadka, S.; Idris, A.M. Receptor model-based source apportionment and ecological risk of metals in sediments of an urban river in Bangladesh. *J. Hazard. Mater.* **2022**, *423*, 127030. [CrossRef] [PubMed]
53. MacDonald, D.D.; Ingersoll, C.G.; Berger, T.A. Development and evaluation of consensus-based sediment quality guidelines for freshwater ecosystems. *Arch. Environ. Contam. Toxicol.* **2000**, *39*, 20–31. [CrossRef] [PubMed]
54. Pandey, M.; Kumar, A.; Mishra, A.; Tripathi, B.D. Application of chemometric analysis and self-Organizing Map-Artificial Neural Network as source receptor modeling for metal speciation in river sediment. *Environ. Pollut.* **2015**, *204*, 64–73. [CrossRef] [PubMed]
55. El-Saadani, Z.; Mingqi, W.; He, Z.; Hamukwaya, S.L.; Abdel Wahed, M.S.M.; Abu Khatita, A. Environmental Geochemistry and Fractionation of Cadmium Metal in Surficial Bottom Sediments and Water of the Nile River, Egypt. *Toxics* **2022**, *10*, 221. [CrossRef] [PubMed]
56. Resongles, E.; Casiot, C.; Freydier, R.; Dezileau, L.; Viers, J.; Elbaz-Poulichet, F. Persisting impact of historical mining activity to metal (Pb, Zn, Cd, Tl, Hg) and metalloid (As, Sb) enrichment in sediments of the Gardon River, Southern France. *Sci. Total Environ.* **2014**, *481*, 509–521. [CrossRef]
57. Atibu, E.K.; Devarajan, N.; Laffite, A.; Giuliani, G.; Salumu, J.A.; Muteb, R.C.; Mulaji, C.K.; Otamonga, J.-P.; Elongo, V.; Mpiana, P.T.; et al. Assessment of trace metal and rare earth elements contamination in rivers around abandoned and active mine areas. The case of Lubumbashi River and Tshamilemba Canal, Katanga, Democratic Republic of the Congo. *Geochemistry* **2016**, *76*, 353–362. [CrossRef]

58. USEPA. Environmental Protection Agency. *Screening Level Ecological Risk Assessment Protocol for Hazardous Waste Combustion Facilities*; USEPA. Environmental Protection Agency: Washington, DC, USA, 1999; Volume 3. Available online: https://www.epa.gov/chemical-research (accessed on 9 October 2021).
59. Ali, M.M.; Ali, M.L.; Islam, M.S.; Rahman, M.Z. Preliminary assessment of heavy metals in water and sediment of Karnaphuli River, Bangladesh. *Environ. Nanotechnol. Moni. Mang.* **2016**, *5*, 27–35. [CrossRef]
60. Raphael, E.C.; Augustina, O.C.; Frank, E.O. Trace metals distribution in fish tissues, bottom sediments and water from Okumeshi River in Delta State, Nigeria. *Environ. Res. J.* **2011**, *5*, 6–10. [CrossRef]
61. Mohiuddin, K.M.; Alam, M.M.; Ahmed, I.; Chowdhury, A.K. Heavy metal pollution load in sediment samples of the Buriganga river in Bangladesh. *J. Bangladesh Agril. Univ.* **2015**, *13*, 229–238. [CrossRef]
62. Ma, X.; Zuo, H.; Tian, M.; Zhang, L.; Meng, J.; Zhou, X.; Liu, Y. Assessment of heavy metals contamination in sediments from three adjacent regions of the Yellow River using metal chemical fractions and multivariate analysis techniques. *Chemosphere* **2016**, *144*, 264–272. [CrossRef] [PubMed]
63. Huffman, A.M.; Sikder, A.M. Assessment OF heavy metal pollution IN the sediments OF the roanoke river. *Water Air Soil Pollut.* **2017**, 229. [CrossRef]
64. Rodríguez-Espinosa, P.F.; Shruti, V.C.; Jonathan, M.P.; Martinez-Tavera, E. Metal concentrations and their potential ecological risks in fluvial sediments of Atoyac River basin, Central Mexico: Volcanic and anthropogenic influences. *Ecotoxicol. Environ. Saf.* **2018**, *148*, 1020–1033. [CrossRef]
65. SEPAC (State Environmental Protection Administration of China). *The Background Values of Soil Elements of China*; Chinese Environmental Science Press: Beijing, China, 1990. (In Chinese)
66. Turekian, K.K.; Wedepohl, K.H. Distribution of the elements in some major units of the earth's crust. *Geol. Soc. Am. Bulletin* **1961**, *72*, 175–192. [CrossRef]
67. Siddique, M.A.M.; Rahman, M.; Rahman, S.M.A.; Hassan, M.R.; Fardous, Z.; Chowdhury, M.A.Z.; Hossain, M.B. Assessment of heavy metal contamination in the surficial sediments from the lower Meghna River estuary, Noakhali coast, Bangladesh. *Int. J. Sediment Res.* **2021**, *36*, 384–391. [CrossRef]
68. Kabir, M.H.; Islam, M.S.; Hoq, M.E.; Tusher, T.R.; Islam, M.S. Appraisal of heavy metal contamination in sediments of the Shitalakhya River in Bangladesh using pollution indices, geo-spatial, and multivariate statistical analysis. *Arab. J. Geosci.* **2020**, *13*, 1135. [CrossRef]
69. Rahman, M.M.; Hassan, M.M. The Amount of Selected Heavy Metals in Water, Sediments and Fish Species from the Rupsha River, Khulna, Bangladesh. *AJFAR* **2020**, *6*, 1–9. [CrossRef]
70. Bhuyan, M.S.; Bakar, M.A.; Nabi, M.R.U.; Senapathi, V.; Chung, S.Y.; Islam, M.S. Monitoring and assessment of heavy metal contamination in surface water and sediment of the Old Brahmaputra River, Bangladesh. *Appl. Water Sci.* **2019**, *9*, 125. [CrossRef]
71. Kormoker, T.; Proshad, R.; Islam, M.S. Ecological Risk Assessment of Heavy Metals in Sediment of the Louhajang River, Bangladesh. *SF J. Environ. Earth Sci.* **2019**, *2*, 1030.
72. Bhuyan, M.S.; Bakar, M.A. Seasonal variation of heavy metals in water and sediments in the Halda River, Chittagong, Bangladesh. *Environ. Sci. Pollut. Res.* **2017**, *24*, 27587–27600. [CrossRef]
73. Hassan, M.; Mirza, A.T.M.; Rahman, T.; Saha, B.; Kamal, A.K.I. Status of heavy metals in water and sediment of the Meghna River, Bangladesh. *Am. J. Environ. Sci.* **2015**, *11*, 427–439. [CrossRef]
74. Islam, M.S.; Ahmed, M.K.; Raknuzzaman, M.; Mamun, M.H.A.; Islam, M.K. Heavy metal pollution in surface water and sediment: A preliminary assessment of an urban river in a developing country. *Ecol. Indic.* **2015**, *48*, 282–291. [CrossRef]
75. Islam, M.M.; Rahman, S.L.; Ahmed, S.U.; Haque, M.K.I. Biochemical characteristics and accumulation of heavy metals in fishes, water and sediments of the river Buriganga and Shitalakhya of Bangladesh. *J. Asian Sci. Res.* **2014**, *4*, 270–279.
76. Rahman, M.S.; Saha, N.; Molla, A.H. Potential ecological risk assessment of heavy metal contamination in sediment and water body around Dhaka export processing zone, Bangladesh. *Environ. Earth Sci.* **2014**, *71*, 2293–2308. [CrossRef]
77. Banu, Z.; Chowdhury, M.S.A.; Hossain, M.D.; Nakagami, K. Contamination and ecological risk assessment of heavy metal in the sediment of Turag River, Bangladesh: An index analysis approach. *J. Water Resour. Prot.* **2013**, *5*, 239–248. [CrossRef]
78. Jolly, Y.N.; Akter, S.; Kabir, J.; Islam, A.; Akbar, S. Trace elements contamination in the river Padma. *Bangladesh J. Phys.* **2013**, *13*, 95–102.
79. Ahmed, A.T.A.; Mandal, S.; Chowdhury, D.A.; Tareq, A.R.M.; Rahman, M.M. Bioaccumulation of some heavy metals in Ayre fish (Sperata Aor Hamilton 1822), sediment and water of Dhaleshwari River in dry season. *Bangladesh J. Zool.* **2012**, *40*, 147–153. [CrossRef]
80. Rashid, H.; Hasan, M.N.; Tanu, M.B.; Parveen, R.; Sukhan, Z.P.; Rahman, M.S.; Mahmud, Y. Heavy metal pollution and chemical profile of Khiru River, Bangladesh. *Int. J. Environ.* **2012**, *2*, 57–63.
81. Mohiuddin, K.M.; Ogawa, Y.; Zakir, H.M.; Otomo, K.; Shikazono, N. Heavy metals contamination in water and sediments of an urban river in a developing country. *Int. J. Environ. Sci. Technol.* **2011**, *8*, 723–736. [CrossRef]

82. Saha, P.K.; Hossain, M.D. Assessment of heavy metal contamination and sediment quality in the Buriganga river, Bangladesh. In *Proceedings of the 2nd International Conference on Environmental Science and Technology, Singapore, 26–28 February, 2011; (EST' 11)*; IACSIT Press: Singapore, 2011; pp. 384–388.
83. Ahmed, M.K.; Islam, S.; Rahman, S.; Haque, M.R.; Islam, M.M. Heavy metals in water, sediment and some fishes of Buriganga River, Bangladesh. *Int. J. Environ. Res.* **2010**, *4*, 321–332.
84. Barhoumi, B.; Beldean-Galea, M.S.; Al-Rawabdeh, A.M.; Roba, C.; Martonos, I.M.; Balc, R.; Kahlaoui, M.; Touil, S.; Tedetti, M.; Driss, M.R.; et al. Occurrence, distribution and ecological risk of trace metals and organic pollutants in surface sediments from a Southeastern European river (Somes¸u Mic River, Romania). *Sci. Total Environ.* **2019**, *660*, 660–676. [CrossRef] [PubMed]
85. Prabakaran, K.; Nagarajan, R.; Eswaramoorthi, S.; Anandkumar, A.; Franco, F.M. Environmental significance and geochemical speciation of trace elements in Lower Baram River sediments. *Chemosphere* **2019**, *219*, 933–953. [CrossRef]
86. Nguyen, B.T.; Do, D.D.; Nguyen, T.X.; Nguyen, V.N.; Nguyen, D.T.P.; Nguyes, M.H.; Truong, H.T.T.; Dong, H.P.; Le, A.H.; Bach, Q.-V. Seasonal, spatial variation, and pollution sources of heavy metals in the sediment of the Saigon River, Vietnam. *Environ. Pollut.* **2020**, *256*, 113412. [CrossRef] [PubMed]
87. Lundy, L.; Alves, L.; Revitt, M.; Wildeboer, D. Metalwater-sediment interactions and impacts on an urban ecosystem. *Int. J. Environ. Res. Publ. Health* **2017**, *14*, 722. [CrossRef] [PubMed]
88. Pehlivan, R. The effect of weathering in the Buyukmelen river basin on the geochemistry of suspended and bed sediments and the hydrogeochemical characteristics of river water, Duzce. Turkey. *J. Asian Earth Sci.* **2010**, *39*, 62–75. [CrossRef]
89. Jones, G.; Jordan, M. The distribution of organic material and trace metals in sediments from the river Liffey estuary, Dublin. *Estuar. Coast. Mar. Sci.* **1979**, *8*, 37–47. [CrossRef]
90. Haque, M.A.; Jewel, M.A.S.; Hasan, J.; Islam, M.M.; Ahmed, S.; Alam, L. Seasonal variation and ecological risk assessment of heavy metal contamination in surface waters of the Ganges River (northwestern Bangladesh). *Malay. J. Analyt. Sci.* **2019**, *23*, 300–311. [CrossRef]
91. Yi, Y.; Yang, Z.; Zhang, S. Ecological risk assessment of heavy metals in sediment and human health risk assessment of heavy metals in fi shes in the middle and lower reaches of the Yangtze River basin. *Environ. Pollut.* **2011**, *159*, 2575–2585. [CrossRef]
92. Proshad, R.; Zhang, D.; Idris, A.M.; Islam, M.S.; Kormoker, T.; Sarker, M.N.I.; Khadka, S.; Islam, A.S.M. Comprehensive evaluation of chemical properties and toxic metals in the surface water of Louhajang River, Bangladesh. Springer. *Environ. Sci. Pollut. Res.* **2021**, *28*, 49191–49205. [CrossRef]
93. Sarkar, M.; Islam, J.B.; Akter, S. Pollution and ecological risk assessment for the environmentally impacted Turag River. Bangladesh. *J. Mater. Environ. Sci.* **2016**, *7*, 2295–2304.
94. Irin, A.; Islam, M.S.; Kabir, M.H.; Hoq, M.E. Heavy metal contamination in water and fishes from the Shitalakhya river at Narayanganj, Bangladesh. *Bangladesh J. Zool.* **2016**, *44*, 267–273. [CrossRef]
95. Islam, M.A.S.; Hossain, M.E.; Majed, N. Assessment of physicochemical properties and comparative pollution status of the Dhaleshwari River in Bangladesh. *Earth* **2021**, *2*, 696–714. [CrossRef]
96. Rahman, M.S.; Ahmed, A.S.S.; Rahman, M.M.; Babu, S.M.O.F.; Sultana, S.; Sarker, S.I.; Awual, R.; Rahman, M.M.; Rahman, M. Temporal assessment of heavy metal concentration and surface water quality representing the public health evaluation from the Meghna River estuary, Bangladesh. Springer. *Appl. Water Sci.* **2021**, *11*, 121. [CrossRef]
97. Choudhury, T.R.; Acter, T.; Uddin, N.; Kamal, M.; Chowdhury, A.M.S.; Rahman, M.S. Heavy metals contamination of river water and sediments in the mangrove forest ecosystems in Bangladesh: A consequence of oil spill incident. *Environ. Nanotechnol. Monit. Manag.* **2021**, *16*, 100484. [CrossRef]
98. Haque, M.M.; Niloy, N.M.; Nayna, O.K.; Fatema, K.J.; Quraishi, S.B.; Park, J.-H.; Kim, K.-W.; Tareq, S.M. Variability of water quality and metal pollution index in the Ganges River, Bangladesh. *Springer. Environ. Sci. Pollut. Res.* **2020**, *27*, 42582–42599. [CrossRef]
99. Proshad, R.; Islam, M.S.; Kormoker, T.; Masud, M.E.M.; Ali, M.M. Assessment of toxic metals contamination with ecological risk of surface water and sediment of Korotoa river in Bangladesh. *Int. J. Adv. Geosci.* **2018**, *6*, 214–221. [CrossRef]
100. Islam, J.B.; Akter, S.; Bhowmick, A.C.; Uddin, M.N.; Sarkar, M. Hydroenvironmental pollution of Turag River in Bangladesh. *Bangladesh J. Sci. Ind. Res.* **2018**, *53*, 161–168. [CrossRef]
101. Shil, S.C.; Islam, M.S.; Irin, A.; Tusher, T.R.; Hoq, M.E. Heavy metal contamination in water and sediments of Passur River near the Sundarbans Mangrove of Bangladesh. *J. Environ. Sci. Nat. Resour.* **2017**, *10*, 15–19. [CrossRef]
102. Bhuiyan, M.A.H.; Samuel Dampare, B.; Islam, M.A.; Suzuki, S. Source apportionment and pollution evaluation of heavy metals in water and sediments of Buriganga River, Bangladesh, using multivariate analysis and pollution evaluation indices. *Environ. Monit. Assess.* **2015**, *187*, 4075. [CrossRef] [PubMed]
103. Hasan, S.J.; Tanu, M.B.; Haidar, M.I.; Ahmed, T.; Rubel, A.K.M.S.A. Physicochemical characteristics and accumulation of heavy metals in water and sediments of the river Dakatia. Bangladesh. *Int. J. Fisher. Aquat. Stud.* **2015**, *2*, 300–304.
104. Rahman, M.M.; Asaduzzaman, M.; Naidu, R. Consumption of arsenic and other elements from vegetables and drinking water from an arsenic-contaminated area of Bangladesh. *J. Hazard. Mater.* **2013**, *262*, 1056–1063. [CrossRef]

105. Hossen, M.A.; Mostafa, M.G. Assessment of heavy metal pollution in surface water of Bangladesh. *Environ. Chall.* **2023**, *13*, 100783. [CrossRef]
106. WHO (World Health Organization). *Guidelines for Drinking Water Quality*, 3rd ed.; World Health Organization: Geneva, Switzerland, 2004; p. 515.
107. Hashim, R.; Song, T.H.; Muslim, N.Z.M.; Yen, T.P. Determination of Heavy Metal Levels in Fishes from the Lower Reach of the Kelantan River, Kelantan, Malaysia. *Trop. Life Sci. Res.* **2014**, *25*, 21–39. [PubMed]
108. Khan, K.; Zeb, M.; Younas, M.; Sharif, H.M.A.; Yaseen, M.; Al-Sehemi, A.G.; Kavil, Y.N.; Shah, N.S.; Cao, X.; Maryam, A.; et al. Heavy metals in five commonly consumed fish species from River Swat, Pakistan, and their implications for human health using multiple risk assessment approaches. *Mar. Pollut. Bulletin.* **2023**, *195*, 115460. [CrossRef] [PubMed]
109. Wahiduzzaman, M.; Islam, M.M.; Sikder, A.H.F.; Parveen, Z. Bioaccumulation and Heavy Metal Contamination in Fish Species of the Dhaleswari River of Bangladesh and Related Human Health Implications. *Biol. Trace Elem. Res.* **2022**, *200*, 3854–3866. [CrossRef] [PubMed]
110. Ahmed, A.S.S.; Rahman, M.; Sultana, S.; Babu, S.M.O.F.; Sarker, M.S.I. Bioaccumulation and heavy metal concentration in tissues of some commercial fishes from the Meghna River estuary in Bangladesh and human health implications. *Mar. Pollut. Bull.* **2019**, *145*, 436–447. [CrossRef]
111. Ahmed, M.K.; Shaheen, N.; Islam, M.S.; Al-Mamun, M.H.; Islam, S.; Islam, M.M.; Kundu, G.K.; Bhattacharjee, L. A comprehensive assessment of arsenic in commonly consumed foodstuffs to evaluate the potential health risk in Bangladesh. *Sci. Total Environ.* **2016**, *544*, 125–133. [CrossRef] [PubMed]
112. Ahmed, M.K.; Baki, M.A.; Islam, M.S.; Kundu, G.K.; Al-Mamun, M.H.; Sarkar, S.K.; Hossain, M.M. Human health risk assessment of heavy metals in tropical fish and shellfish collected from the river Buriganga, Bangladesh. *Environ. Sci. Pollut. Res.* **2015**, *22*, 15880–15890. [CrossRef]
113. Rahman, M.S.; Molla, A.H.; Saha, N.; Rahman, A. Study on heavy metals levels and its risk assessment in some edible fishes from Bangshi River, Savar, Dhaka, Bangladesh. *Food Chem.* **2012**, *134*, 1847–1854. [CrossRef]
114. USFDA. *Food and Drug Administration. Guidance Document for Nickel in Shellfish*; DHHS/PHS/FDA/CFSAN/; Office of Seafood: Washington, DC, USA, 1993. Available online: http://www.cfsan.fda.gov/~frf/guid-as.html (accessed on 17 March 2016).
115. FAO. Compilation of legal limits for hazardous substances in fish and fishery products. *FAO Fish Circular* **1983**, *464*, 5–100.
116. WHO. Environmental Health Criteria No 165: Inorganic Lead. World Health Organization (WHO); 1995. Available online: http://www.inchem.org/documents/ehc/ehc/ehc165.htm (accessed on 26 December 2024).
117. FAO/WHO. (FAO/World Health Organization). *Codex Alimentarius—General Standards for Contaminants and Toxins in Food*; Schedule 1 maximum and guideline levels for contaminants and toxins in food. Reference CX/FAC 02/16; Joint FAO/WHO Food Standards Programme, Codex Committee: Rotterdam, The Netherlands, 2002.
118. IOM. *Institute of Medicine Dietary Reference Intakes. Applications in Dietary Planning. Subcommittee on Interpretation and Uses of Dietary Reference Intakes and the Standing Committee on the Scientific Evaluation of Dietary Reference Intakes*; Institute of Medicine of the National Academies, The National Academies Press: Washington, DC, USA, 2003; p. 248.
119. EC. European Commission of the European Communities. Commission regulation (EC) n. 221/2002 of the 6 February 2002 amending regulation (EC) n. 466/2002 setting maximum levels for certain contaminants in foodstuffs. *Off. J. Eur. Commun.* **2002**, *7*, 37.
120. Lee, H.J.; Park, M.K.; Seo, Y.R. Pathogenic mechanisms of heavy metal-induced Alzheimer's disease. *Toxicol. Environ. Health Sci.* **2018**, *10*, 1–10. [CrossRef]
121. Wise, J.P., Jr.; Young, J.L.; Cai, J.; Cai, L. Current understanding of hexavalent chromium [Cr (VI)] neurotoxicity and new perspectives. *Environ. Int.* **2022**, *158*, 106877. [CrossRef] [PubMed]
122. Chen, A.; Kim, S.S.; Chung, E.; Dietrich, K.N. Thyroid hormones in relation to lead, mercury, and cadmium exposure in the National Health and Nutrition Examination Survey, 2007–2008. *Environ. Health Perspect.* **2013**, *121*, 181–186. [CrossRef]
123. Yang, A.-M.; Lo, K.; Zheng, T.-Z.; Yang, J.-L.; Bai, Y.-N.; Feng, Y.-Q.; Cheng, N.; Liu, S.-M. Environmental heavy metals and cardiovascular diseases: Status and future direction. *Chronic Dis. Transl. Med.* **2020**, *6*, 251–259. [CrossRef]
124. Patwa, J.; Flora, S.J.S. Heavy metal-induced cerebral small vessel disease: Insights into molecular mechanisms and possible reversal strategies. *Int. J. Mol. Sci.* **2020**, *21*, 3862. [CrossRef]
125. Capitão, C.; Martins, R.; Santos, O.; Bicho, M.; Szigeti, T.; Katsonouri, A.; Bocca, B.; Ruggieri, F.; Wasowicz, W.; Tolonen, H.; et al. Exposure to heavy metals and red blood cell parameters in children: A systematic review of observational studies. *Front. Pediatr.* **2022**, *10*, 921239. [CrossRef] [PubMed]
126. Wilbur, S.; Abadin, H.; Fay, M.; Yu, D.; Tencza, B.; Ingerman, L.; Klotzbach, J.; James, S. *Toxicological Profile for Chromium*; Agency for Toxic Substances and Disease Registry: Atlanta, GA, USA, 2012.
127. Hu, X.; Fernandes, J.; Jones, D.P.; Go, Y.M. Cadmium stimulates myofibroblast differentiation and mouse lung fibrosis. *Toxicology* **2017**, *383*, 50–56. [CrossRef] [PubMed]

128. Ma, Y.; Ran, D.; Shi, X.; Zhao, H.; Liu, Z. Cadmium toxicity: A role in bone cell function and teeth development. *Sci. Total Environ.* **2021**, *769*, 144646. [CrossRef]
129. Kim, D.-W.; Ock, J.; Moon, K.-W.; Park, C.-H. Association between Pb, Cd, and Hg exposure and liver injury among Korean adults. *Int. J. Environ. Res. Pub.* **2021**, *18*, 6783. [CrossRef] [PubMed]
130. Balali-Mood, M.; Naseri, K.; Tahergorabi, Z.; Khazdair, M.R.; Sadeghi, M. Toxic mechanisms of five heavy metals: Mercury, lead, chromium, cadmium, and arsenic. *Front. Pharmacol.* **2021**, *12*, 643972. [CrossRef]
131. Mandal, B.K.; Suzuki, K.T. Arsenic round the world: A review. *Talanta* **2002**, *58*, 201. [CrossRef]
132. Lei, H.-L.; Wei, H.-J.; Ho, H.-Y.; Liao, K.-W.; Chien, L.-C. Relationship between risk factors for infertility in women and lead, cadmium, and arsenic blood levels: A cross-sectional study from Taiwan. *BMC Public Health* **2015**, *15*, 1220–1311. [CrossRef] [PubMed]
133. Lin, J.; Lin, X.; Qiu, J.; You, X.; Xu, J. Association between heavy metals exposure and infertility among American women aged 20–44 years: A cross-sectional analysis from 2013 to 2018 NHANES data. *Front. Public Health* **2023**, *11*, 1122183. [CrossRef] [PubMed]
134. NYSDOH. Hopewell Precision Area Contamination: Appendix C-NYS DOH. In *Procedure for Evaluating Potential Health Risks for Contaminants of Concern*; New York State Department of Health: New York, NY, USA, 2007.
135. Adebiyi, F.M.; Ore, O.T.; Owolafe, O.S. Human health risk assessment of potentially toxic metals in fish (*Cynoglossus* sp.) commonly consumed in Nigeria. *Discov. Toxicol.* **2024**, *1*, 7. [CrossRef]
136. Yi, Y.; Tang, C.; Yi, T.; Yang, Z.; Zhang, S. Health risk assessment of heavy metals in fish and accumulation patterns in food web in the upper Yangtze River, China. *Ecotoxicol. Environ. Saf.* **2017**, *145*, 295–302. [CrossRef]
137. Hossain, M.M.; Wang, S.; Liang, Z.; Geng, A.; Jahan, I.; Tripty, S.J.; Maxwell, S.J.; Hossain, I.; Sethupathy, S.; Zhu, D. Potentially toxic elements (PTEs) in the invasive Asian clam (*Corbicula fluminea*) from polluted urban river areas of Bangladesh and evaluation of human health risk. *Environ. Monit. Assess.* **2024**, *196*, 1178. [CrossRef] [PubMed]
138. Anyanwu, B.O.; Chris, D.I. Human health hazard implications of heavy metals concentration in swimming crab (*Callinectes amnicola*) from polluted creeks in Rivers State, Nigeria. *Case Stud. Chem. Environ. Eng.* **2023**, *7*, 100325. [CrossRef]
139. Li, P.; Huang, Y.-Y.; Zeng, J.; Lin, Z.Z.; Huang, Z.-Y. Health risk assessment of heavy metals in shellfish collected from Fujian, China. *Hum. Ecol. Risk Assess. Int. J.* **2020**, *26*, 621–635. [CrossRef]
140. Yu, B.; Wang, X.; Dong, K.F.; Xiao, G.; Ma, D. Heavy metal concentrations in aquatic organisms (fishes, shrimp and crabs) and health risk assessment in China. *Mar. Pollut. Bull.* **2020**, *159*, 1115. [CrossRef] [PubMed]
141. Jiang, H.; Qin, D.; Chen, Z.; Tang, S.; Bai, S.; Mou, Z. Heavy Metal Levels in Fish from Heilongjiang River and Potential Health Risk Assessment. *Bull. Environ. Contam. Toxicol.* **2016**, *97*, 536–542. [CrossRef]
142. Chai, M.; Li, R.; Gong, Y. Bioaccessibility-corrected health risk of heavy metal exposure via shellfish consumption in coastal region of China. *Environ. Pollut.* **2021**, *273*, 116529. [CrossRef]
143. Hasanein, S.S.; Mourad, M.H.; Haredi, A.M.M. The health risk assessment of heavy metals to human health through the consumption of Tilapia spp and catfish caught from Lake Mariut, Egypt. *Heliyon* **2022**, *8*, e09807. [CrossRef] [PubMed]
144. Kaba, P.; Shushi, S.; Gyimah, E.; Husein, M.; Abomohra, A. Multivariate Analysis of Heavy Metals and Human Health Risk Implications Associated with Fish Consumption from the Yangtze River in Zhenjiang City, China. *Water* **2023**, *15*, 1999. [CrossRef]

toxics

MDPI

Article

Copper-Induced Neurodegenerative Disorders and Therapeutic Potential of Curcumin-Loaded Nanoemulsion

Govind Hake [1], Akshada Mhaske [1], Rahul Shukla [1,*] and Swaran Jeet Singh Flora [2,*]

[1] Department of Pharmaceutics, National Institute of Pharmaceutical Education and Research-Raebareli, Near CRPF Base Camp, Bijnor-Sisendi Road, Sarojini Nagar, Lucknow 226002, India; govindkh.765mspe21@niperrbl.ac.in (G.H.)

[2] Era College of Pharmacy, Era Lucknow Medical University, Sarfarajgunj, Lucknow Hardoi Road, Lucknow 226002, India

* Correspondence: rahul.shukla@niperraebareli.edu.in (R.S.); swaran.jeet@erauniversity.in or sjsflora@hotmail.com (S.J.S.F.)

Abstract: Copper accumulation in neurons induces oxidative stress, disrupts mitochondrial activity, and accelerates neuronal death, which is central to the pathophysiology of neurodegenerative diseases like Wilson disease. Standard treatments for copper toxicity, such as D-penicillamine, trientine, and chloroquine, are frequently associated with severe side effects, creating a need for safer therapeutic alternatives. To address this, we developed a curcumin-loaded nanoemulsion (CUR-NE) using the spontaneous emulsification technique, aimed at enhancing the bioavailability and therapeutic efficacy of curcumin. The optimized nanoemulsion displayed a particle size of 76.42 nm, a zeta potential of −20.4 mV, and a high encapsulation efficiency of 93.69%, with a stable and uniform structure. The in vitro tests on SH-SY5Y neuroblastoma cells demonstrated that CUR-NE effectively protected against copper-induced toxicity, promoting significant cellular uptake. Pharmacokinetic studies revealed that CUR-NE exhibited a longer half-life and extended circulation time compared to free curcumin. Additionally, pharmacodynamic evaluations, including biochemical assays and histopathological analysis, confirmed that CUR-NE provided superior neuroprotection in copper overload conditions. These results emphasize the ability of CUR-NE to augment the therapeutic effects of curcumin, presenting a novel approach for managing copper-induced neurodegeneration. The study highlights the effectiveness of nanoemulsion-based delivery platforms in improving chelation treatments for neurological diseases.

Keywords: copper toxicity; Wilson's disease; mitochondrial oxidative stress; chelation therapy; neurotoxicity; herbal molecules

1. Introduction

Copper (Cu) toxicity occurs when excess copper accumulates in the body due to impaired regulation, especially in the liver and brain. While Cu is essential for biological functions, excessive amounts lead to toxicity. This is often caused by defects in Cu-transporting proteins like ATP7A or ATP7B [1,2]. In the brain, excess Cu generates reactive oxygen species (ROS), leading to oxidative damage to lipids, proteins, and DNA, which damages neurons. Cu also disrupts mitochondrial function, reducing energy production and exacerbating oxidative stress, particularly in regions like the basal ganglia, impairing motor control and cognition [3]. Cu accumulation activates glial cells, triggering neuroinflammation, which worsens neuronal damage. The combined effects of oxidative stress, mitochondrial dysfunction, and inflammation impair cellular repair, contributing to neurodegeneration [4]. First-line treatment for Cu toxicity typically involves the use of metal chelation therapy and zinc supplementation. Metal chelators such as D-penicillamine, dimercaptosuccinic acid, and trientine are used to bind excess Cu, facilitating its excretion through urine. Concurrently, zinc salts are administered to reduce Cu absorption in the gastrointestinal tract, limiting its systemic circulation [5,6]. These combined strategies effectively manage Cu overload and prevent further toxic effects on organs such as the liver and brain [7]. Synthetic chelators used to treat Cu toxicity can lead to several adverse effects, such as kidney damage, gastrointestinal discomfort, and bone marrow suppression. Additionally, they necessitate careful monitoring due to potential Cu deficiency and interactions with other treatments. Zinc therapy, although beneficial, may be less effective in severe cases and can cause side effects with prolonged use [8,9]. Herbal chelators present a promising alternative, offering a safer profile with fewer side effects. These natural agents may help modulate Cu levels and combat oxidative damage, providing a viable long-term treatment option with a reduced risk of adverse effects [10]. Curcumin (CUR) is a natural bioactive polyphenol obtained by the extraction process from rhizomes of *Curcuma longa* L. [11]. Numerous studies have reported that CUR exhibits antioxidant, anticancer, anti-inflammatory, antimicrobial, antiepileptic, antidepressant, immunomodulatory, neuroprotective, antiapoptotic, and antiproliferative properties [12]. It acts as an ROS scavenger, increasing the glutathione (GSH) level by inducing the glutathione cysteine ligase. CUR may protect the brain from Aβ toxicity in Alzheimer's disease animal models by chelating copper sulfate (Cu^{2+}) ions [13]. With its various applications, the most common problems concerning the biopharmaceuticals of CUR are poor aqueous solubility, instability, rapid metabolism by phase II reaction in the hepatocytes with biliary excretion, and poor intestinal permeability, so it is challenging to incorporate into aqueous formulations [14].

The clinical applications of CUR are diverse, with promising potential in various therapeutic areas. In cancer treatment, CUR's ability to bind Cu^{2+} has been linked to its effectiveness in limiting tumor growth and preventing angiogenesis [15]. It reduces inflammation by chelating metals and suppressing pathways like NF-κB. It also boosts antioxidant defenses by activating key enzymes such as catalase and superoxide dismutase [16]. Furthermore, CUR has shown applicability in neutralizing the harmful effects of heavy metals like lead, mercury, and cadmium by mitigating oxidative damage. However, its broader clinical use faces obstacles, such as limited absorption in the body, non-selective metal-binding that may affect essential minerals like iron and zinc, and the need for higher doses that can lead to gastrointestinal discomfort [17]. Additionally, CUR's interactions with certain medications, particularly those involved in blood clotting or acid regulation, require careful consideration. Although early-stage research provides encouraging results, more comprehensive clinical studies are needed to validate its therapeutic potential, underscoring the importance of developing methods to enhance its absorption and overall efficacy [18]. To address the physicochemical challenges associated with CUR, lipid-based encapsulation technologies, such as nanoemulsion, could be explored. Nanoemulsion can efficiently solubilize large amounts of lipophilic compounds within oil droplets, which are stabilized by a surfactant film at the interface [19]. In this study, we focused on developing a stable CUR-loaded nanoemulsion and assessed its effectiveness in counteracting Cu^{2+}-induced neurotoxicity in a rodent model [20,21]. CUR was encapsulated in nanoemulsion to enhance its solubility, stability, and bioavailability, overcoming its poor water solubility and rapid metabolism. The nano-sized droplets improve absorption, cellular penetration, and targeted delivery, boosting its efficacy for clinical use. Ginger oil nanoemulsion, rich in antioxidant compounds like 6-shogaol and gingerol, combat metal toxicity by reducing oxidative stress. The nanoemulsion technology enhances bioavailability, stability, and targeted delivery, offering a promising natural solution for managing metal-induced damage.

2. Result and Discussion

2.1. Pre-Formulation Analysis

2.1.1. Solubility Evaluation of Curcumin

The solubility of CUR in various components of the formulation, including the oil phase, surfactants, and cosurfactants, was assessed. The analysis was performed using ginger oil as the primary oil phase, along with diethylene glycol monoethyl ether (DGME), Tween 80, ethanol, PEG 400, and Cremophor EL. The solubility values observed for CUR in these components were 1.47 ± 0.91 mg/mL in ginger oil, 15.3 ± 1.68 mg/mL in DGME, 8.95 ± 0.81 mg/mL in Tween 80, 5.29 ± 1.74 mg/mL in ethanol, 20.0 ± 1.55 mg/mL in PEG 400, and 4.13 ± 1.21 mg/mL in Cremophor EL [22]. The solubility profile is depicted in Figure S1A,B.

2.1.2. Selection of Oil, Surfactant, and Co-Surfactant

The oil selection was based on solubility and miscibility of CUR in oil. Ginger oil was preferred based on its solubility with other excipients and antioxidant activity in various neurodegenerative disorders [23]. Tween 80 and polyethylene glycol 400 (PEG 400) surfactants were selected in preliminary studies. Transcutol P and ethanol (EoH) were selected as a cosurfactant due to the property of permeation enhancement [24]. It was noted that ginger oil was miscible in Tween 40, 60, and 80; Transcutol P; Cremophor EL; and ethanol, and it was immiscible with a span of 20, 80, and PEG 400. Based on the evaluation, the above-mentioned surfactants and cosurfactants were selected for preliminary formulation batches.

2.1.3. Design of Pseudo-Ternary Phase Diagram

Pseudo-ternary phase diagrams were developed using ginger oil as the oil phase, with surfactant mixture of Tween 80 combined with ethanol, PEG 400 and DGME as cosurfactants, and water as the aqueous phase. Nanoemulsion prepared with a S_{mix} of PEG 400 and ethanol exhibited a biphasic appearance. Adjusting the PEG 400-to-ethanol did not significantly expand the clear nanoemulsion region. Increasing the ginger oil concentration from 5% to 15% (w/w) substantially reduced the transparent nanoemulsion zone, indicating a concentration-dependent effect. Increasing surfactant levels, along with PEG 400 and ethanol, led to a larger clear nanoemulsion region. Excessive surfactant concentrations have been associated with safety concerns due to potential toxicity. Among the tested formulations, the combination of Tween 80 and ethanol produced the clearest nanoemulsion, as depicted in Figure S2. In contrast, increasing the ethanol ratio to 1:2 reduced the nanoemulsion region, likely due to a decrease in the interfacial barrier between the oil and aqueous phases. Varying the Tween 80 ratio showed no significant changes in the clear region, which could be attributed to the inherent phase behaviour of the ternary system. Nanoemulsion formulated with Tween 80 and DGME showed slightly smaller clear regions compared to the Tween 80 and ethanol system. Nanoemulsions are kinetically stable systems, formed at precise concentrations of oil, surfactant, cosurfactant, and water, and exhibit no phase separation, creaming, or cracking. Formulations identified from the phase diagrams underwent stress stability testing, including centrifugation, freeze–thaw cycles, and heating–cooling cycles. Additionally, temperature fluctuations during testing disrupt stability by causing phase separation and altering droplet distribution due to curvature free energy changes. Combinations that showed no evidence of phase separation, creaming, cracking, coalescence, or phase inversion during stress stability tests were deemed stable and selected for further studies, as summarized in Table 1.

Table 1. Thermodynamic stability testing of NE region obtained from pseudo-ternary phase diagram using Tween 80 and ethanol (1:1) (+ indicates stability, whereas − indicates instability).

Batch No.	% v/v of Solvent			Heating and Cooling Cycle	Centrifugation	Freeze–Thaw Cycle
	Oil	Smix	Water			
A1	5	25	70	+	−	−
A2	5	30	65	+	+	−
A3	5	35	60	+	+	+
A4	5	40	55	+	+	+
A5	5	45	50	+	+	+
A6	5	50	45	+	+	+
A7	5	55	40	+	−	−
A8	5	60	35	−	−	−
A9	5	65	30	+	+	+
A10	5	70	25	−	+	−
A11	10	40	50	+	−	−
A12	10	45	45	+	+	+
A13	10	50	40	+	+	+
A14	10	55	35	−	−	−
A15	10	60	30	−	+	+
A16	10	65	25	+	+	−
A17	15	55	30	+	−	−
A18	15	60	25	+	+	+
A19	15	65	20	+	−	−
A20	15	70	15	+	+	+

2.2. Characterization of CUR-NE Formulation

The particle size of CUR-loaded nanoemulsion (CUR-NE) plays a crucial role in optimizing its oral bioavailability, as smaller particles can enhance drug absorption and improve permeation across the intestinal barrier. The particle size, polydispersity index, and zeta potential of the CUR-NE were measured using a Zetasizer. The optimized formulations results showed a hydrodynamic diameter of 76.42 nm with a PDI of 0.266, indicating a narrow and uniform particle size distribution. The zeta potential, which provides insight into the stability of the formulation, was at −20.34 mV, suggesting that the CUR-NE is stable. This negative zeta potential is likely due to the anionic nature of CUR. To further examine the particle morphology, scanning electron microscopy (SEM) was used, revealing that the particles were spherical, evenly spaced, and mostly smaller than 100 nm. The particle size obtained from the Zetasizer was notably smaller than that observed in the SEM analysis [19]. The percentage drug content in the nanocarrier was found to be 93.69%, indicating efficient encapsulation of CUR within the oil droplets. FTIR analysis showed distinct functional group peaks for each excipient. In the CUR-NE, the intensity of CUR-specific peaks was reduced, likely due to its encapsulation in the internal phase of the nanoemulsion. Figure 1A displays the FTIR peaks for CUR, excipients, and the formulation. Ketone stretching (-C=O) was observed at 1542.65 cm^{-1} in CUR and at 1645.26 cm^{-1} in CUR-loaded nanoemulsion. Aromatic -C-H bending appeared at 883.44 cm^{-1} in ginger oil and at 878.62 cm^{-1} in CUR-loaded nanoemulsion. The alcoholic -O-H bending was observed at 1044.99 cm^{-1} in ethanol and at 1045.24 cm^{-1} in the nanoemulsion. The ether -C-O stretching was seen at 1095.36 cm^{-1} in Tween 80 and at 1085.54 cm^{-1} in CUR-loaded nanoemulsion. These observations confirm that all characteristic peaks of the excipients were present in the CUR-NE. The FTIR spectra of CUR, ginger oil, Tween 80, ethanol, and the CUR-loaded nanoemulsion are shown in Figure 1A. The SEM images revealed that the CUR-NE particles were spherical, with a smooth surface morphology. Minimal aggregation of the particles was observed, which can be attributed to the surfactant concentration used in the formulation (Figure 1B).

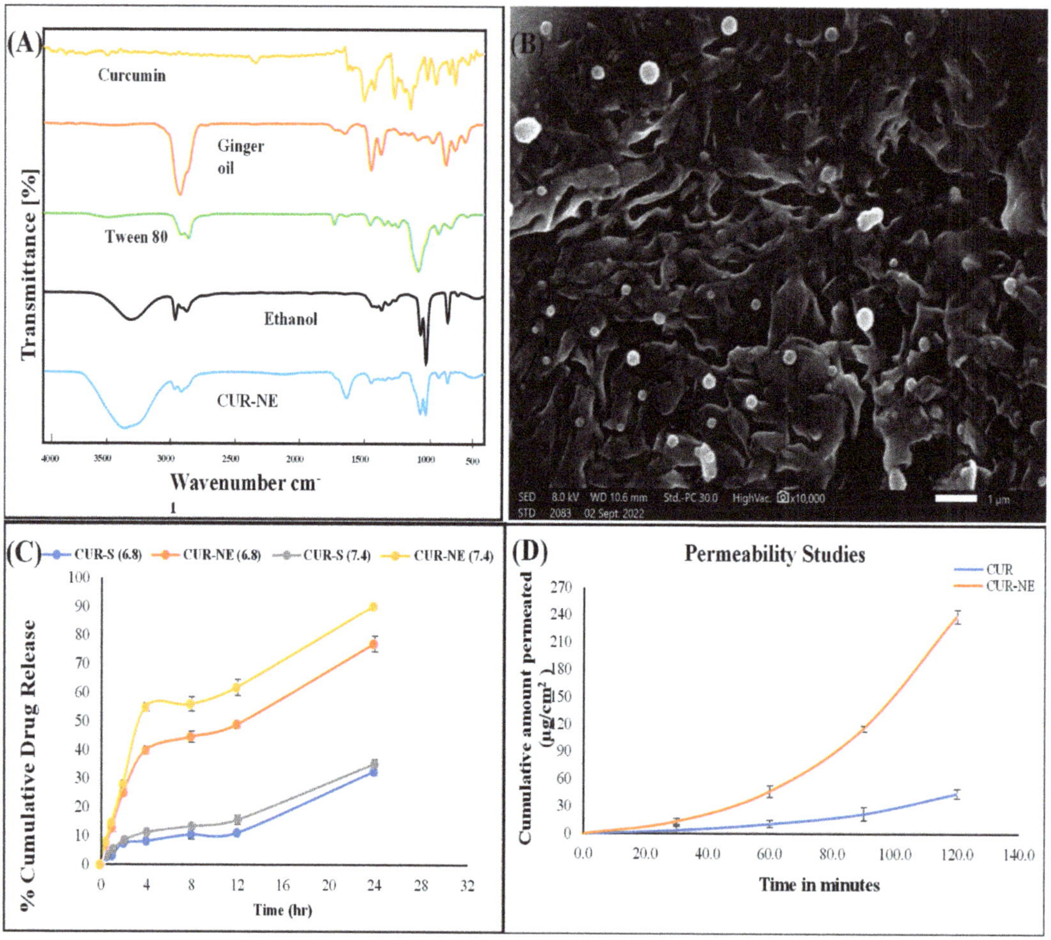

Figure 1. (**A**) FTIR spectra of curcumin, ginger oil, Tween 80, ethanol, and CUR-NE. (**B**) SEM micrograph of CUR-NE showing spherical, small, rounded particles. (**C**) Cumulative drug release (%) from CUR suspension and CUR-NE in phosphate buffer at pH 6.8 and 7.4, with all samples evaluated in triplicate. (**D**) The permeability of the prepared nanoformulation was assessed using the intestinal sac model, with all samples analysed in triplicate. 1: 1 is part of unit of wavenumber. It should read as cm^{-1}.

2.3. *In Vitro Release Profile Analysis*

The in vitro drug release behaviour of the CUR solution and optimized nanoemulsion was assessed using the dialysis bag method across different physiological pH levels (6.8, and 7.4). After 2 h, the cumulative CUR release was 7.46%, and 8.56% for the drug solution, while for the CUR -loaded nanocarriers, it was 25.00% and 27.99%, respectively, in (6.8 and 7.4) buffer. These findings demonstrate that CUR release from the nanocarriers was notably faster than from the drug solution. The release data are depicted in Figure 1C. This increased release rate can be attributed to the presence of surfactants and cosurfactants in the formulation, which reduce interfacial barriers and facilitate drug diffusion into the release medium. Release kinetics analysis further confirmed that the CUR release from the nanocarriers followed a first-order kinetic model [25].

2.4. Ex Vivo Intestinal Permeability Studies

The ex vivo intestinal permeation of free CUR and CUR-NE was evaluated using the non-everted intestinal gut sac technique. At 2 h, the cumulative permeation of CUR per unit area was recorded, with free CUR reaching 44.26 $\mu g/cm^2$, while CUR-NE demonstrated a significantly higher permeation of 238.96 $\mu g/cm^2$. These results indicate that CUR-NE exhibited a significantly higher absorption compared to free CUR at various time intervals. The cumulative permeation of CUR over time is illustrated in Figure 1D.

2.5. Thermodynamic Stability Studies

The physicochemical stability of the prepared nano-formulation was evaluated by monitoring changes in particle size, polydispersity index (PDI), and zeta potential over a one-month period at 4 °C, 25 °C, and 45 °C. The results showed that the CUR-NE formulation remained stable under different storage temperature conditions. The stability outcome is shown in Table 2.

Table 2. Stability data.

Sampling Time	Size (nm) ± SD	PDI ± SD	ZP (mV) ±SD
At 4 °C			
Initial	76.42 ± 0.09	0.192 ± 0.79	−20.5 ± 0.75
After 10 days	76.42 ± 0.09	0.245 ± 0.67	−20.2 ± 0.43
After 20 days	76.12 ± 0.45	0.217 ± 1.56	−19.5 ± 0.48
After 30 days	75.56 ± 0.79	0.282 ± 0.71	−18.1 ± 1.18
At Room Temperature			
Initial	76.42 ± 0.11	0.266 ± 0.43	−20.4 ± 0.47
After 10 days	76.50 ± 0.74	0.200 ± 0.76	−20.4 ± 0.51
After 20 days	77.19 ± 0.81	0.268 ± 0.44	−20.3 ± 0.93
After 30 days	76.98 ± 1.23	0.289 ± 0.21	−20.0 ± 0.71
At 45 °C			
Initial	76.42 ± 0.29	0.200 ± 0.09	−20.7 ± 0.86
After 10 days	77.30 ± 0.76	0.219 ± 0.37	−20.3 ± 0.73
After 20 days	77.45 ± 0.83	0.242 ± 0.80	−19.4 ± 0.69
After 30 days	78.41 ± 1.44	0.282 ± 0.91	−18.19 ± 0.85

2.6. Cell Line Study

2.6.1. Cell Viability Analysis

The SH-SY5Y cells were exposed to varying concentrations of Cu^{2+} (ranging from 1 to 50 μM) to identify the most suitable dose for subsequent experiments. Cu^{2+} treatment resulted in a concentration-dependent decrease in cell viability, with a 50 μM concentration showing a significant reduction of 49.98%. Based on these results, 50 μM Cu^{2+} was chosen to induce neurotoxicity in further studies. Additional tests revealed that treatment with equivalent CUR suspension and CUR-NE, at concentrations of 2.5 μM and 5 μM, effectively mitigated the Cu^{2+}-induced toxicity and enhanced cell viability. The cell viability analysis is represented in Figure 2A,B.

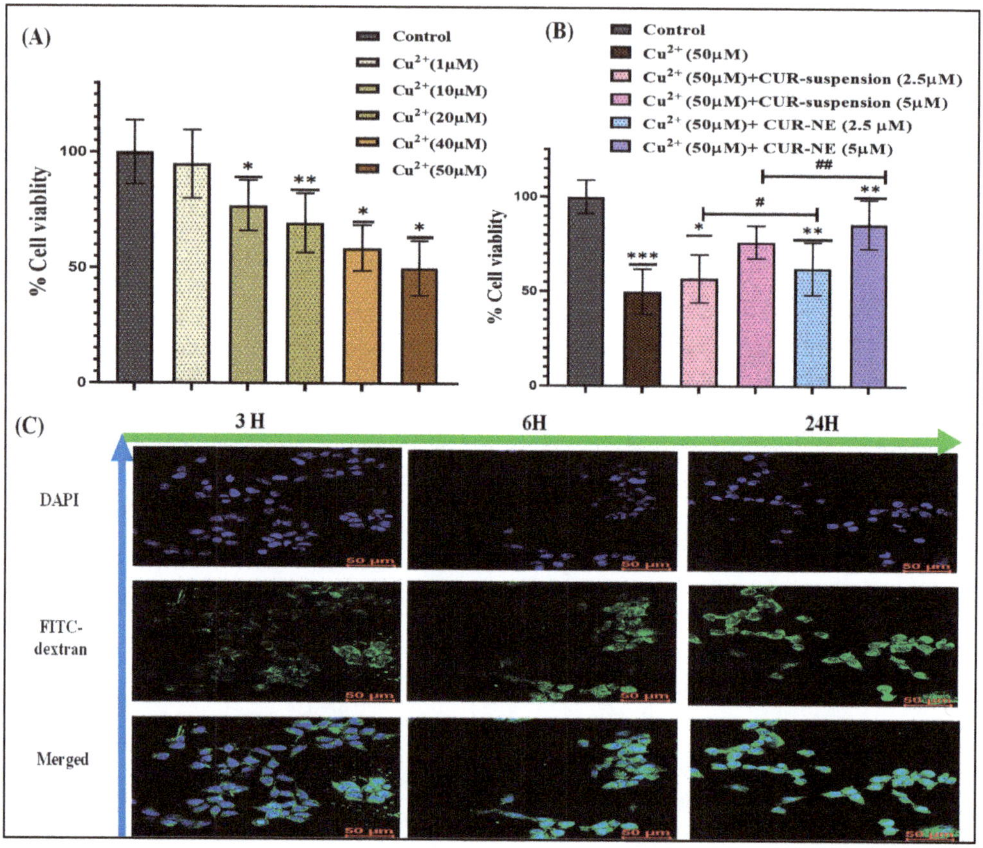

Figure 2. (**A**) Cell viability analysis of Cu^{2+} at concentrations ranging from 1 to 50 μM. (**B**) Cell viability analysis of Cu^{2+} (50 μM) with various treatment groups. (**C**) Confocal microscopy images illustrating the cellular uptake of FITC-loaded nanoemulsion in SH-SY5Y cells at 3, 6, and 24 h incubation time points (40× magnification). Data were analyzed using one-way ANOVA with Dunnett's multiple comparison. All experiments were conducted in triplicates (n = 3). Data were analyzed using one-way ANOVA with Dunnett's multiple comparison. All treatment groups were statistically compared to the control group to assess differences in the measured outcomes (denoted with *, *** $0.0001 < p \leq 0.001$, ** $0.001 < p \leq 0.01$, * $p \leq 0.05$). The CUR vs. CUR-NE groups were statistically analyzed using a one-way ANOVA followed by the Sidak multiple comparisons test (denoted with #). ##: $0.001 < p \leq 0.01$, #: $p \leq 0.05$.

2.6.2. Cellular Uptake Evaluation

The uptake of the FITC-labeled formulation in SH-SY5Y cells was monitored at 3, 6, and 24 h. The analysis showed a gradual, time-dependent increase, with the highest uptake occurring at 24 h, a moderate level at 6 h, and the lowest at 3 h. This pattern indicates progressive accumulation of the FITC-labeled nanoformulation within the cells as the incubation time increased. The uptake dynamics are presented in Figure 2C.

2.7. *In Vivo Evaluation*

2.7.1. Pharmacokinetic Assessment

The investigation aimed to compare the pharmacokinetics and tissue distribution of CUR-NE with CUR suspension following oral administration. Samples from plasma, brain, liver, and kidneys were collected to evaluate the absorption, distribution, and elimination profiles of both regimens. CUR-NE showed a significantly higher brain concentration, reaching 1229.64 ± 88.99 ng/g, compared to the suspension. In plasma, CUR-NE achieved a peak concentration of 1207.35 ± 79.14 ng/mL, while the suspension displayed a higher value of 1507.33 ± 84.11 ng/mL. Liver concentrations for the suspension peaked at 1392.163 ± 91.12 ng/g, whereas CUR-NE reached a maximum of 921.04 ± 75.33 ng/g. Similarly, kidney concentrations were highest for the suspension at 1391.53 ± 52.99 ng/g, whereas CUR-NE reached 736.53 ± 61.69 ng/g.

The AUC_{Brain} for CUR-NE was significantly higher 16,330.34 ng/mg·h, in contrast to 2719.98 ng/g·h for the suspension, suggesting a marked improvement in bioavailability. In terms of brain concentration, CUR-NE achieved a peak value at (T_{max}) of 4 h, while the suspension reached only 9 h, indicating superior brain targeting. Furthermore, the brain targeting index for CUR-NE was 4.7, underscoring its enhanced ability to effectively accumulate in brain tissues. These findings are further detailed in Figure 3A–D and Table 3.

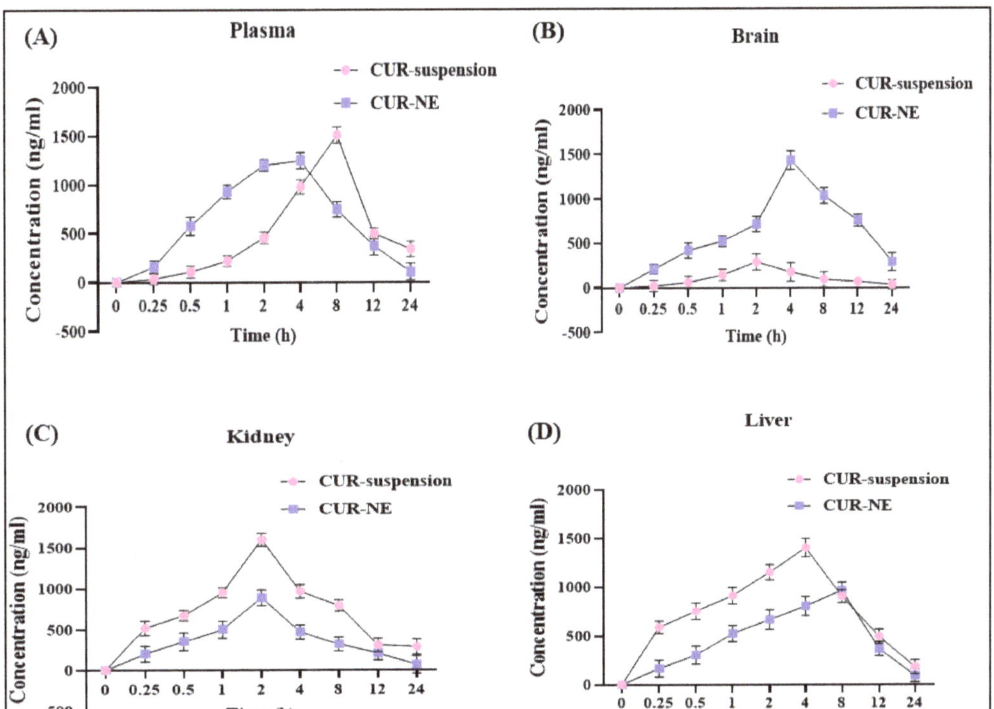

Figure 3. (**A**) Pharmacokinetic estimation of CUR in plasma at various time points. (**B**) Pharmacokinetic estimation of CUR in the brain at different time points. (**C**) Pharmacokinetic estimation of CUR in the kidney at multiple time points. (**D**) Pharmacokinetic estimation of CUR in the liver at several time points. All experiments were conducted with a sample size of n = 6.

Table 3. Pharmacokinetic parameters and CUR quantification in plasma and brain.

Parameter	Units	Plasma		Brain	
		CUR Suspension	CUR-NE	CUR Suspension	CUR-NE
t1/2	1/h	5.63	8.71	6.5	9.99
Tmax	h	8	4	9	4
Cmax	h	1507.33	1207.35	373.92	1229.64
AUC 0-t	ng/mL·h	15,865.64	12,055.48	2719.98	16,330.34
MRT 0-inf_obs	h	15.56	8.73	8.99	11.35

2.7.2. Pharmacodynamic Evaluation

The Morris water maze (MWM) test was conducted to compare the effects of orally administered CUR-NE with those of CUR suspension. The parameters assessed included the time taken to locate the platform and the duration spent in the target zone. These results were analysed across the control group, Cu^{2+}-treated group, and Cu^{2+}-treated groups with treatments groups. The study showed that rats receiving CUR-NE exhibited memory retention comparable to the control group, whereas Cu^{2+}-treated rats demonstrated impaired memory. While rats treated with CUR suspension also showed improved memory, the effect was less significant compared to those receiving CUR-NE (Figure 4A,B).

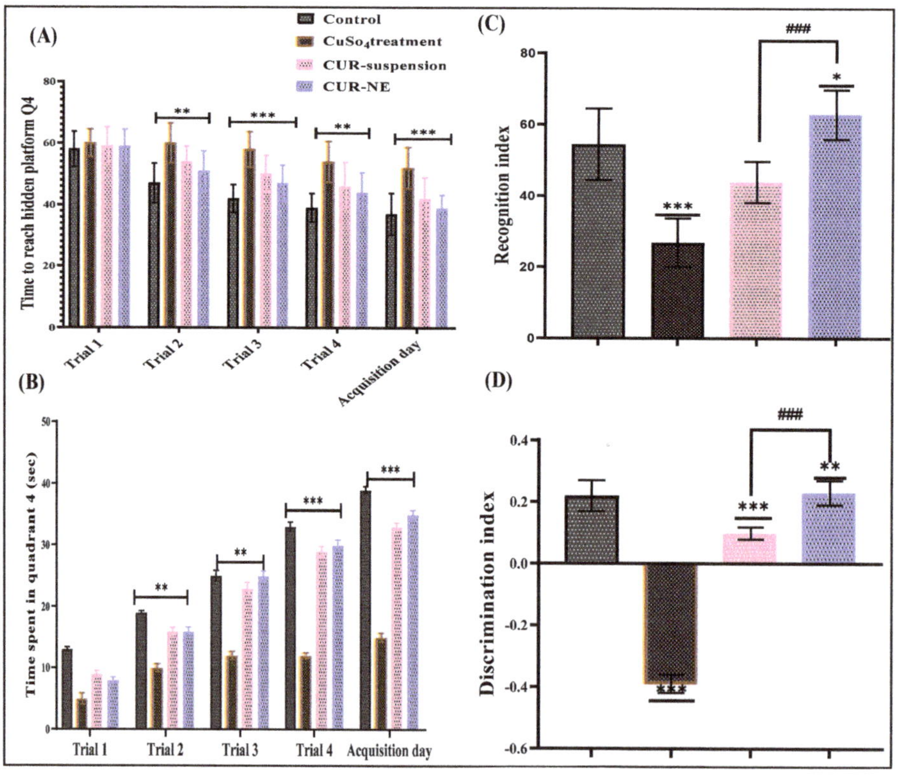

Figure 4. (A) Morris's water maze (MWM) test showing total time to reach the platform in Q4 for control, disease control, Cu+ CUR, and Cu+ CUR-NE groups (n = 5). (B) Graphical representation of

the time spent in Q4 ($n = 5$). (**C**) Recognition index for different treatment groups. (**D**) Discrimination index for different treatment groups. Statistical analysis for panels (**A**,**B**) was performed using two-way ANOVA with Dunnett's multiple comparison; all groups were compared with control. All experiments were conducted with a sample size of $n = 6$. For (**C**,**D**) graph, data were analyzed using one-way ANOVA with Dunnett's multiple comparison. All treatment groups were statistically compared to the control group to assess differences in the measured outcomes (denoted with *, *** $0.0001 < p \leq 0.001$, ** $0.001 < p \leq 0.01$, * $p \leq 0.05$). The CUR vs. CUR-NE groups were statistically analyzed using a one-way ANOVA followed by the Sidak multiple comparisons test (denoted with ###); ### $0.0001 < p \leq 0.001$.

Results from the Novel Object Recognition Test (NORT) are presented in Figure 4C,D. During the probe test (R2), rats exposed to Cu^{2+} displayed a marked decrease in preference for the novel object compared to the control group. Conversely, rats treated with Cu^{2+} + CUR and Cu^{2+} + CUR-NE showed significant improvements in novel object preference compared to the Cu^{2+} group. In the treatment groups (CUR and CUR-NE during T2), a strong preference for the novel object was observed, indicating notable cognitive improvement compared to the Cu^{2+}-treated group.

2.7.3. Biochemical Estimation of Neuronal Oxidative Stress Markers

Superoxide dismutase (SOD) is crucial for the scavenging of reactive oxygen species (ROS). The highest SOD activity was observed in the control group, followed by the Cu^{2+} + CUR-NE and Cu^{2+} + CUR treatment groups. The lowest SOD activity was found in the group exposed to Cu^{2+} alone [26]. The graphical representation of SOD estimation is represented in Figure 5A.

Catalase (CAT) activity changes serve as an indicator of oxidative stress levels. The highest catalase activity was observed in the control group, followed by the Cu^{2+} + CUR-NE and Cu^{2+} + CUR groups. In contrast, the group exposed to Cu^{2+} alone showed the lowest catalase activity. The catalase activity data are illustrated in Figure 5B. In the malondialdehyde (MDA) assessment, treatment with CUR-NE and CUR reduced MDA levels compared to the Cu^{2+} group, as shown in Figure 5C. Nitrite (NO) levels, which indicate the production of nitrogen species that can damage cellular structures, were elevated in the Cu^{2+}-induced neurotoxicity group compared to the other groups. Figure 5D shows that treatment with CUR and CUR-NE effectively reduced excessive nitrite levels, offering potential protection against oxidative stress caused by Cu^{2+} toxicity. In neurodegenerative diseases, oxidative stress results from ROS-induced damage to neurons, potentially leading to their degeneration and dysfunction. The elevated ROS levels in Cu^{2+}-induced rats indicate a significant increase in oxidative stress, highlighting the neurotoxic effects of Cu^{2+} [27]. The Cu^{2+} + CUR and Cu^{2+} + CUR-NE groups exhibited a decrease in ROS activity, indicating that CUR and CUR-NE may promote the chelation of Cu^{2+}. The ROS measurement results are depicted in Figure 5E. The molecule 8-hydroxy-2′-deoxyguanosine (8-OHdG) is a prominent marker of oxidative stress, specifically indicating DNA damage. In the case of in vivo Cu^{2+} -induced neurotoxicity, higher levels of 8-OHdG signify an increase in oxidative DNA damage compared to the disease + treatment groups, as illustrated in Figure 5F. The results, as shown in Figure S3, indicate that prolonged exposure to Cu^{2+} for 16 weeks caused an increase in brain Cu^{2+} levels compared to the control group. Treatment with CUR suspension and CUR-NE demonstrated its chelation efficacy by significantly reducing Cu^{2+} concentrations in brain tissue, highlighting its potential therapeutic effect.

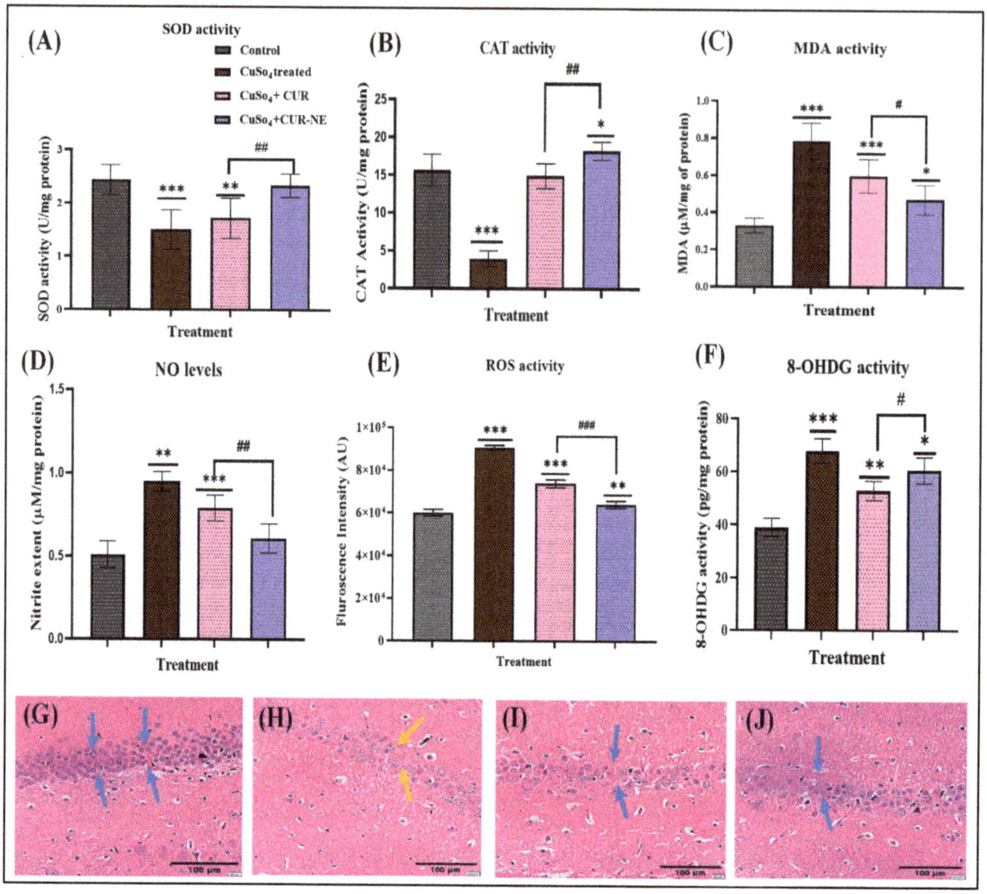

Figure 5. (**A**) Evaluation of superoxide dismutase (SOD) activity. (**B**) Evaluation of catalase (CAT) activity. (**C**) Measurement of malondialdehyde (MDA) levels. (**D**) Assessment of nitric oxide (NO) activity. (**E**) Determination of reactive oxygen species (ROS) levels. (**F**) Evaluation of 8-hydroxy-2′-deoxyguanosine (8-OHdG) levels. (**G**) Histopathological analysis of brain tissue in the control group. (**H**) Histopathological analysis of brain tissue in the Cu^{2+} control group. (**I**) Histopathological evaluation of brain tissue in the Cu^{2+} + CUR suspension group. (**J**) Histopathological analysis of brain tissue in the Cu^{2+} + CUR-NE group. All experiments were conducted with a sample size of $n = 6$. Data were analyzed using one-way ANOVA with Dunnett's multiple comparison. All treatment groups were statistically compared to the control group to assess differences in the measured outcomes (denoted with *, *** $0.0001 < p \leq 0.001$, ** $0.001 < p \leq 0.01$, * $p \leq 0.05$). The CUR vs. CUR-NE groups were statistically analyzed using a one-way ANOVA followed by the Sidak multiple comparisons test (denoted with #); ###: $0.0001 < p \leq 0.001$, ##: $0.001 < p \leq 0.01$, #: $p \leq 0.05$.

2.8. Histopathological Evaluation

In the brain histopathological assessment, the control group showed no signs of inflammation or cellular imbalance. In contrast, the diseased groups exhibited significant brain structural changes, including cellular swelling, vacuolation, myelin breakdown, and neuronal damage [28]. The Disease + CUR/CUR-NE groups displayed mild alterations in the myelin sheath, with fewer cellular abnormalities (Figure 5G–J). Importantly, both the CUR suspension and CUR-NE groups demonstrated healthier brain tissue compared to the Cu^{2+}-treated rats [29]. The blue arrows indicate the absence of inflammation or cellular imbalance, representing healthy morphology in the control group. In contrast, the yellow arrows highlight signs of neuronal damage and structural alterations, reflecting pathological changes.

3. Materials and Method

Curcumin (99% purity) was purchased from Himedia Laboratory, India. Ginger oil was purchased from Sigma, Germany. Tween 80, PEG 400, and Transcutol P were purchased from SRL, Taloja. Ethanol and methanol were purchased from Merk, Mumbai. The 6-diamidino-2-phenylindole (DAPI) and FITC-dextran were purchased from Sigma-Aldrich (USA). The 3-(4,5-Dimethylthiazol-2-yl)-2,5-Diphenyltetrazolium bromide) (MTT), Dulbecco's Modified Eagle Medium (DMEM), Tris-HCl buffer, fetal bovine serum, and penicillin-streptomycin (penstrep) antibiotics were obtained from Thermo Scientific™. Triple-distilled water, used in the formulation preparation, was sourced from the Milli-Q system (Millipore, Merck) and was in-house. Biochemical assay kits were purchased from Bioassay Technology Laboratory.

3.1. Pre-Formulation Studies

3.1.1. Solubility Analysis of Curcumin

The solubility of CUR was evaluated in various oils (olive, clove, and ginger oil), surfactants (Tween 40, 60, 80, and Span series), and cosurfactants (PEG 400, ethanol, Cremophor EL, Transcutol P, and soy lecithin). CUR was added to 1.5 mL of each selected oil, surfactant, and cosurfactant mixture in microcentrifuge tubes. The tubes were vortexed using a Spinix vortex mixer to ensure complete dispersion. To achieve equilibrium, the tubes were incubated at 37 ± 1.0 °C in a shaking water bath for 72 h [30]. After completion of 72 h, the samples were diluted in methanol and quantified using UV spectroscopy at 425 nm [31].

3.1.2. Design of Pseudo-Ternary Phase Diagram

The pseudo-ternary phase diagram assists in the understanding of the spontaneous emulsification process. The pseudo-ternary point diagram demonstrated clear, turbid, and biphasic states, which were coded with colors. Ginger oil, Tween 80 as a surfactant, Transcutol P, and ethanol as a cosurfactant was used in NEs components after the assessments of different oils, surfactants, and cosurfactants on the basis of miscibility and solubility analysis. The pseudo-ternary phase diagrams were designed using oil, S_{mix}, and distilled water and were built using the aqueous titration method. For the selected Tween 80 and DGME, ethanol was mixed in different ratios of 1:1, 1:2, and 2:1. Each oil: S_{mix} ratio was slowly titrated with a water phase, and visual inspection of the samples was done for indications of separation of the phases [32].

3.2. Preparation of Curcumin-Loaded Nanoemulsion (CUR-NE)

CUR-loaded nanocarriers were synthesized using a spontaneous emulsification technique, which involved the gradual addition of an aqueous phase into a pre-mixed oil and surfactant blend under controlled conditions with gentle stirring to form stable O/W nanoemulsions. The emulsification process is governed by factors such as phase transition regions, interfacial tension, viscosity, structural properties, and the concentration of surfactants. The preparation method was executed in three stages: initially, an organic solution was prepared by dissolving 4 mg of CUR in 5% ginger oil. Subsequently, a surfactant system comprising hydrophilic surfactant (Tween 80) and cosurfactant (ethanol) was formulated. Finally, the aqueous phase was introduced dropwise into the CUR-containing oily phase while stirring continuously at room temperature to achieve uniform emulsification. This method yielded CUR-encapsulated nanocarriers with desirable characteristics for potential applications [33].

Thermodynamic Stability Analysis

A thermodynamic stability study was accomplished for the various batches of nanoemulsion manufactured by varying the concentration of stabilizer and water chosen from the pseudo-ternary phase diagrams at different stress conditions [34]. The stability of the samples was analyzed through three different tests: the heating–cooling cycle, centrifugation, and freeze–thaw cycle. The heating–cooling cycle test involved subjecting the formulations to six temperature series ranging from 4 °C to 45 °C, with each temperature maintained for at least 48 h. Following this, a centrifugation test was conducted on the formulations to evaluate their stability and identify those that remained stable under these conditions. To evaluate the stability of the samples, batches were first centrifuged at 3500 rpm for 30 min. Samples that showed no signs of instability after centrifugation were then subjected to a freeze–thaw cycle. This test involved subjecting the samples to repeated freezing and thawing to check for any phase separation or instability under temperature changes. Three consecutive freeze–thaw cycles were conducted at temperatures ranging from −20 °C to room temperature over a 48 h period for the storage of samples. Only those samples that remained clear and did not exhibit any phase separation were selected for further analysis.

3.3. Characterization of CUR-NE

3.3.1. Droplet Size and Surface Charge Analysis

The droplet diameter, PDI, and zeta potential were assessed using Zetasizer (Malvern Instruments Ltd., Worcester, UK). A total of 200 μL test samples of nanocarriers were mixed with 2 mL triple-distilled water, and the measurement was taken in triplicate by using normal disposable cuvettes for size and disposable folded capillary cuvette for zeta potential at 25 °C temperature [34].

3.3.2. Drug Contents

The nanoformulation was subjected to centrifugation at 1200 rpm, and the resulting pellet was resuspended in methanol. The drug content was determined using HPLC (Agilent 1100 series, (Agilent Technologies, Memphis, TN, USA) at 425 nm. The percentage of drug content in the CUR-NE was calculated using the following formula:

$$\%\ \text{Drug contents} = \frac{\text{Amt. of CUR measured by Spectrophotomer}}{\text{Total amount of CUR added}} * 100$$

3.3.3. FTIR Analysis

The compatibility of excipients and the purity of the formulation were assessed using FT-infrared spectroscopy (Bruker Alpha-P FTIR, Bruker, Boston, MA, USA). A sample consisting of the drug and excipients was placed on the sample holder, and the clean probe was positioned accordingly. The analysis was performed over a wavelength range of 450 to 4000 cm^{-1} to evaluate the interactions and determine the purity of the formulation [35]. The obtained data were evaluated on OPUS software version 9.0 and plotted.

3.3.4. Morphological Characterization of Prepared Nanocarrier

Scanning electron microscopy (SEM, JEOL JSM-IT 200, Tokyo, Japan) was employed to assess the droplet size and surface morphology of the nanoemulsion. A 15 μL aliquot of the sample was placed on a stub covered with carbon tape and allowed to dry in a vacuum desiccator for 3 days. Prior to imaging, the sample was gold-coated for 2 min to enhance conductivity. The prepared sample was then mounted onto the sample holder, and its morphology was examined under low voltage using the SEM [36].

3.3.5. Stability Studies

The accelerated stability of the CUR-NE was assessed by storing the formulation at temperatures of 45 °C, 25 °C, and 4 °C for one month. Particle size, PDI, and zeta potential were measured using dynamic light scattering (DLS) at regular time intervals to track any changes in the formulation's stability over time under different storage conditions [37].

3.3.6. In Vitro Drug Release Studies

The release profiles of CUR and CUR-NE were assessed using a dialysis bag with a molecular weight cutoff of 12,000 Dalton. The study was conducted in three different pH conditions (pH 1.2, 6.8, and 7.4) to simulate physiological environments. The dialysis bag was pre-soaked overnight in the release medium to hydrate the membrane. A 2 mL aliquot of CUR and CUR-NE (equivalent to 4 mg) was placed in the dialysis bag and submerged in 100 mL of release medium, which was continuously stirred at 100 rpm and maintained at 37 ± 0.5 °C. At specified time intervals (0.5, 1, 2, 4, 8, 12, and 24 h), 2 mL samples were withdrawn from the release medium and replaced with fresh medium to ensure sink conditions. The drug concentration was measured using UV–visible spectrophotometry at 425 nm [38,39].

3.3.7. Permeability Studies

The intestinal permeability of free drug and nanoformulation was evaluated by utilizing the non-everted intestinal (rats) sac technique. The intestine was taken from sacrificed SD rat of the control group from an animal laboratory. The small intestine was freed from intestinal content by passing the oxygenated cold normal saline solution with the blunt syringe. Cleaned intestinal segments were cut into 8 cm long pieces and placed into a Krebs Ringers buffer (7.4 pH) solution with oxygenation. A blunt needle was used to inject each sac with an equivalent concentration of CUR suspension and CUR-NE, and both sides of the intestine were then securely tied with silk thread. This non-everted intestine was placed in beakers having 100 mL of Krebs Ringer buffer solution (7.4 pH) on a magnetic stirrer with 100 rpm at 37 °C temperature equipped with laboratory aerators. An amount of 5 mL of aliquot was taken at programmed time intervals and replaced with the same volume of KRB to keep the sink state [40]. The samples were investigated by using Agilent 1100 series HPLC, (Memphis, TN, USA) techniques at 425 nm at room temperature. The cumulative amount permeated per unit area (μg/cm^2) of intestinal sacs was assessed.

3.4. Cell Line Protocol

The SH-SY5Y cell line was obtained from institutional cell line repository. Cells were seeded in the T25 flask under 5% CO_2 incubator. The media composition consists of equal volumes of Minimum Essential Medium (MEM) and Dulbecco's Modified Eagle Medium (DMEM) supplemented with 0.1% penicillin–streptomycin solution.

3.4.1. Cell Viability Analysis

Cells were cultured in a 96-well plate and allowed to adhere for 12 h. After cell adhesion, the complete medium was substituted with incomplete medium, and treatments were carried out in two distinct experimental setups. In the first setup, cells were exposed to Cu^{2+} at varying concentrations of 1, 10, 20, 40, and 50 μM. In the second setup, cells were subjected to different treatment combinations, including Cu^{2+} at 50 μM alone, Cu^{2+} at 50 μM with CUR suspension at 2.5 μM, Cu^{2+} at 50 μM with CUR suspension at 5 μM, Cu^{2+} at 50 μM with CUR-NE at 2.5 μM, and Cu^{2+} at 50 μM with CUR-NE at 5 μM. After

24 h, the media were replaced with MTT solution, and cell viability was assessed using a multiplate reader (Synergy H1, BioTek, Woburn, MA, USA). Wells treated with 0.1% DMSO served as the 100% viability control, while Cu^{2+} (50 μM) acted as the positive control. The percentage of cell viability was calculated using the following formula. All the readings were calculated in triplicates. Data were analyzed using one-way ANOVA with Dunnett's multiple comparison. All treatment groups were statistically compared to the control group to assess differences in the measured outcomes (denoted with *). The CUR vs. CUR-NE groups were statistically analyzed using a one-way ANOVA followed by the Sidak multiple comparisons test. Cell viability was defined according to the following equation:

$$\textit{Cell viability} = [\textit{Absorbance of treatment group}/\textit{Absorbance of control group}] \times 100$$

3.4.2. Cellular Uptake Analysis

The SH-SY5Y cells were seeded in 12-well plates and allowed to adhere for 12 h. Following the attachment period, the cells were treated with FITC-labeled nanoemulsion (NE) for 3, 6, and 24 h in a CO_2 incubator. After each treatment interval, the cells were stained with DAPI for 5 min, followed by removal of the DAPI solution and washing with fresh PBS. The coverslips were then mounted onto glass slides with 80% glycerol and sealed with clear gel polish. The slides were examined using confocal microscopy (Leica Microsystems DMI8, Wetzlar, Germany), with images captured in both the green and blue channels to assess the cellular uptake and distribution of the NE.

3.5. *Animal Study Protocol*

Sprague Dawley rats (SD) with 180 to 250 g weight range were selected. The animals were acclimatized for 7 days and then were sorted via further randomization for treatment. The pharmacokinetic and pharmacodynamic study (protocol no. NIPER/RBL/IAEC 192 and 8 March 2024) was approved by the Institutional Animal Ethics Committee (IAEC) at NIPER-Raebareli.

3.5.1. Pharmacokinetic Analysis

Following an overnight fast, the rats were given an oral dose of CUR suspension at 80 mg/kg (prepared in 0.5% sodium carboxymethyl cellulose) and an equivalent dose of CUR-NE (at equivalent concentration). The blood collection was carried via retroorbital route under slight anesthesia in the following intervals: 0.25, 0.5, 1, 2, 4, 8, 12, and 24 h in EDTA tubes. After 24 h, the rats were euthanized, and tissues like brain, kidneys, liver, and lungs were isolated. These tissues were rinsed with PBS (pH 7.4), dried, weighed, and homogenized in PBS using a tissue homogenizer. The resulting homogenates were stored at −80 °C till evaluation. The samples were analyzed with HPLC analysis (emodin was used as internal standard at 510 nm and CUR at 425 nm. The mobile phase was composed of 0.05 M sodium acetate buffer: Acetonitrile at 60:40 ratio at 1 mL per minute. The isolated homogenate samples were centrifuged with 500 μL acetonitrile at 12,000 rpm for 5 min at 4 °C. The separated organic phase was evaporated to dryness and reconstituted with acetonitrile and evaluated on HPLC instrument. The obtained data were analyzed with the PKSolver software version 10. All the readings were calculated in (n = 6). Data were analyzed using one-way ANOVA with Dunnett's multiple comparison. All treatment groups were statistically compared with the control group to assess differences in the measured outcomes (denoted with *). The CUR vs. CUR-NE groups were statistically analyzed using a one-way ANOVA followed by the Sidak multiple comparisons test (denoted with #). The drug targeting index was defined according to the following equation:

$$Drug\ targeting\ Index = \frac{(AUC1)B1/(AUC1)p1}{(AUC2)B2/(AUC2)p2}$$

B1 represents the AUC of CUR-NE in brain;
B2 represents the AUC of CUR-suspension in brain;
P1 represents the AUC of CUR-NE in plasma;
P2 represents the AUC of CUR suspension in plasma.

3.5.2. Pharmacodynamic Study

The animals were randomly assigned to four groups. In Group I (control), SD rats received daily intraperitoneal (IP) injections of 0.9% NaCl. Group II was treated with Cu^{2+} (20 mg/kg) daily for 16 weeks in drinking water. In Group III, rats were given Cu^{2+} (20 mg/kg) daily for 16 weeks, followed by 14 days of oral CUR treatment at 80 mg/kg. Group IV received Cu^{2+} (20 mg/kg) daily for 16 weeks, followed by 14 days of oral CUR-NE treatment at 80 mg/kg. Thus, the Cu^{2+} treatment lasted 16 weeks, after which the CUR and CUR-NE treatments were administered for 14 days. All the readings were calculated in (n = 6). Data were analyzed using one-way ANOVA with Dunnett's multiple comparison. All treatment groups were statistically compared to the control group to assess differences in the measured outcomes (denoted with *). The CUR vs. CUR-NE groups were statistically analyzed using a one-way ANOVA followed by the Sidak multiple comparisons test (denoted with #).

(i). Morris water maze test

The Morris water maze (MWM) test was employed to assess learning and memory performance after exposure to Cu^{2+} and subsequent treatments. Five rats from each group participated in a four-day training period, during which they explored various quadrants to locate a hidden platform submerged in the water. The time taken to reach the platform was recorded during each trial. On the fifth day, a probe trial was conducted to evaluate memory retention, measuring the time taken to identify the target quadrant and the duration spent within it. Following the behavioral tests, the rats were sacrificed, and tissue samples were collected for further analysis [41].

(ii). Nobel Object Recognition Test

The Novel Object Recognition Test (NORT) was conducted to evaluate the effects of Cu^{2+} toxicity on memory function. The setup included a black open-top box measuring 65 cm × 65 cm × 45 cm, with a high-definition camera (Lenovo 300 FHD Webcam) placed above to record the rats' behavior. The test was divided into three stages: habituation, familiarization, and recognition. During the habituation stage, rats were allowed to explore the empty box freely. In the familiarization stage, they were introduced to two identical objects placed within the box. Finally, in the recognition phase, one of the familiar objects was replaced with a novel object, and the time spent exploring each object was recorded [42].

3.6. Assessment of Oxidative Stress Biomarkers in the Brain

(i). Quantification of Superoxide Dismutase Quantity

The brain homogenate was prepared by suspending the tissue in Tris-HCl buffer, followed by centrifugation (sigma Z-16pk, Osterode, Germany) to collect the supernatant. For the assay, a reaction mixture was prepared by combining 0.2 mL of the brain supernatant with 0.8 mL of distilled water and 0.2 mL of NADH. After incubation, the reaction was terminated by adding acetic acid. The mixture was allowed to stand for 10 min, and the absorbance was measured at 560 nm using a spectrophotometer to assess the intensity of the reaction.

(ii). Quantification of Catalase Activity

CAT activity in brain tissue was measured by combining the homogenate with 1 mL of substrate solution. The mixture was allowed to incubate for 2 min, after which the reaction was terminated with 1 mL of ammonium molybdate, resulting in the formation of a yellow-colored complex. The intensity of the color was then quantified by measuring the absorbance using a spectrophotometer.

(iii). Quantification of MDA

To quantify thiobarbituric acid-reactive substances (TBARSs) in brain tissue, 100 µL of brain homogenate was combined with 250 µL of 20% acetic acid, 250 µL of thiobarbituric acid, and 8% sodium dodecyl sulfate, followed by the addition of distilled water to reach the desired volume in a test tube. The mixture was incubated at 90 °C for 1 h and then rapidly cooled under running tap water. After cooling, the samples were centrifuged, and the absorbance was measured at 532 nm using a spectrophotometer. The results were compared to a standard curve for quantification.

(iv). Quantification of Nitric oxide Levels

Brain nitric oxide levels were measured using the Griess reagent. To perform the test, 100 µL of the reagent was combined with equivalent volume of brain homogenate. After incubation, the resulting reaction product was analyzed at 540 nm using a spectrophotometer. A standard curve was created using sodium nitrite, and the nitrite levels in the brain samples were calculated and reported in micromoles per milligram of protein.

(v). Estimation of ROS

The DCFDA assay was employed to assess oxidative stress levels in brain tissue. For the assay, brain homogenate was mixed with DCFDA solution and 980 µL of buffer (pH 7.4). The mixture was incubated in the dark for 20 min to prevent light interference. Fluorescence measurements were obtained using a multimode plate reader, with excitation at 483 nm and emission at 530 nm. The oxidative stress levels were quantified as fluorescence units (FU) per milligram of protein.

(vi). Quantification of Neuroinflammatory Marker (8-OHdG)

To assess the levels of 8-hydroxy-2′-deoxyguanosine (8-OHdG) in brain tissue, the homogenate was combined with a commercially available pre-mixed solution and incubated for 20 min. Following incubation, 200 µL of Folin–Ciocalteu reagent was introduced to the mixture, and the incubation continued. The absorbance of the resulting reaction was measured using a spectrophotometer, and the concentration of 8-OHdG was calculated by referencing a standard curve.

(vii). Assessment of cerebral Cu^{2+} content

Brain Cu^{2+} levels were quantified using inductively coupled plasma mass spectrometry (ICP-MS). In summary, brain tissue samples were subjected to a standard acid digestion process. The resulting digested solutions were analysed for Cu^{2+} concentration using a PerkinElmer emission spectrometer.

All the readings were calculated in ($n = 6$). Data were analyzed using one-way ANOVA with Dunnett's multiple comparison. All treatment groups were statistically compared to the control group to assess differences in the measured outcomes (denoted with *). The CUR vs. CUR-NE groups were statistically analyzed using a one-way ANOVA followed by the Sidak multiple comparisons test (denoted with #).

3.7. Histopathological Evaluations

After the behavioural studies, the animals were sacrificed, and their brains were carefully extracted. The tissues were then placed in a fixation solution until they became firm. Following fixation, the brain tissue was embedded in paraffin wax and allowed to solidify in a mold. Once solidified, the tissue was sliced into 5 mm thick sections. These sections underwent a series of alcohol washes before being stained with hematoxylin and eosin. Finally, the stained slides were examined under a microscope for analysis [2,43].

4. Conclusions

In this study, we highlight the significant potential of CUR-NE in overcoming the challenges associated with CUR's therapeutic application in neurodegenerative diseases. Our developed nanoformulation effectively enhances CUR's pharmacokinetic profile and improves its systemic bioavailability, which is crucial for targeting neurological conditions. By amplifying CUR 's anti-inflammatory, antioxidant, and metal chelation properties, this nanoemulsion offers a multifaceted approach to addressing metal-induced neurodegeneration. The inclusion of synergistic agents like ginger oil further strengthens the therapeutic impact, potentially improving outcomes for diseases linked to oxidative stress and metal toxicity. Our findings, including improvements in cognitive function and memory in rat models, suggest the broader applicability of developed formulations in treating Cu^{2+} induced neurodegenerative disorders. We recognize the need for further investigation, particularly regarding long-term safety, optimal dosing, and efficacy in clinical studies. Overall, this work emphasizes the growing role of nanotechnology in enhancing the effectiveness of natural compounds, offering new avenues for the treatment of complex neurological diseases.

Supplementary Materials: The following supporting information can be downloaded at https://www.mdpi.com/article/10.3390/toxics13020108/s1, Figure S1: Solubility analysis: (A) Solubility in co-surfactants (B) Solubility in surfactants. All data were calculated in triplicate, Figure S2: Ternary phase diagram (A) Tween 80: DEME in 1:1, 1:2 and 2:1 ratio, (B) Tween 80: ethanol in 1:1, 1:2 and 2:1 ratio, (C) PEG400: ethanol in 1:1, 1:2 and 2:1 ratio, Figure S3: Evaluation of cerebral copper content in different group (N = 3). All the readings were calculated in triplicates. Data was analyzed using one-way ANOVA with Dunnett's multiple comparison. All treatment groups were statistically compared to the control group to assess differences in the measured outcomes (denoted with *). The CUR vs. CUR-NE groups were statistically analyzed using a one-way ANOVA followed by the Sidak multiple comparisons test.

Author Contributions: Conceptualization, S.J.S.F.; Methodology, A.M. and G.H.; Software, A.M.; Validation, R.S.; Formal analysis, G.H., R.S. and S.J.S.F.; Resources, A.M., G.H. and R.S.; Data curation, A.M., G.H. and R.S.; Writing—original draft, R.S.; Writing—review and editing, S.J.S.F.; Visualization, S.J.S.F.; Supervision, R.S. All authors have read and agreed to the published version of the manuscript.

Funding: The financial support was received from the Department of Pharmaceuticals (DoP), Ministry of Chemicals and Fertilizers, Government of India. The NIPER-R communication number is NIPER-R/Communication/711.

Institutional Review Board Statement: The study was conducted according to the guidelines of the Declaration of Helsinki, and was approved by the Institutional Animal Ethics Committee (IAEC) at NIPER-Raebareli (protocol no. NIPER/RBL/IAEC 192 and 8 March 2024).

Informed Consent Statement: Not applicable.

Data Availability Statement: The data that support the findings of this study are available from the corresponding author upon reasonable request.

Conflicts of Interest: The authors declare no conflict of interest.

References

1. Roberts, E.A. Update on the Diagnosis and Management of Wilson Disease. *Curr. Gastroenterol. Rep.* **2018**, *20*, 56. [CrossRef]
2. La Fontaine, S.; Mercer, J.F.B. Trafficking of the copper-ATPases, ATP7A and ATP7B: Role in copper homeostasis. *Arch. Biochem. Biophys.* **2007**, *463*, 149–167. [CrossRef] [PubMed]
3. Scheiber, I.F.; Brůha, R.; Dušek, P. Pathogenesis of Wilson disease. In *Handbook of Clinical Neurology*; Elsevier: Amsterdam, The Netherlands, 2017; pp. 43–55.
4. Mhaske, A.; Sharma, S.; Shukla, R. Nanotheranostic: The futuristic therapy for copper mediated neurological sequelae. *J. Drug Deliv. Sci. Technol.* **2023**, *80*, 104193. [CrossRef]
5. Litwin, T.; Dzieżyc, K.; Członkowska, A. Wilson disease—Treatment perspectives. *Ann. Transl. Med.* **2019**, *7*, S68. [CrossRef] [PubMed]
6. Llanos, R.M.; Mercer, J.F.B. The molecular basis of copper homeostasis and copper-related disorders. *DNA Cell Biol.* **2002**, *21*, 259–270. [CrossRef]
7. Członkowska, A.; Litwin, T. Wilson disease—Currently used anticopper therapy. In *Handbook of Clinical Neurology*; Elsevier: Amsterdam, The Netherlands, 2017; pp. 181–191.
8. Singh, S.K.; Balendra, V.; A Obaid, A.; Esposto, J.; A Tikhonova, M.; Gautam, N.K.; Poeggeler, B. Copper-mediated β-amyloid toxicity and its chelation therapy in Alzheimer's disease. *Metallomics* **2022**, *14*, mfac018. [CrossRef]
9. Abolaji, A.O.; Fasae, K.D.; Iwezor, C.E.; Farombi, E.O. D-Penicillamine prolongs survival and lessens copper-induced toxicity in Drosophila melanogaster. *Toxicol. Res.* **2020**, *9*, 346–352. [CrossRef] [PubMed]
10. Mhaske, A.; Shukla, R.; Flora, S.J.S. Modulation of copper-induced neurotoxicity by monoisoamyl 2,3-dimercaptosuccinic acid loaded nanoparticles through inhibition of mitophagy and reduction of oxidative stress in SH-SY5Y cells. *Toxicol. Rep.* **2025**, *14*, 101874. [CrossRef]
11. Leung, M.H.M.; Harada, T.; Kee, T.W. Delivery of Curcumin and Medicinal Effects of the Copper(II)-Curcumin Complexes. *Curr. Pharm. Des.* **2013**, *19*, 2070–2083.
12. Pan-On, S.; Dilokthornsakul, P.; Tiyaboonchai, W. Trends in advanced oral drug delivery system for curcumin: A systematic review. *J. Control. Release* **2022**, *348*, 335–345. [CrossRef]
13. Shen, L.; Zhang, H.Y.; Ji, H.F. A theoretical study on Cu(II)-chelating properties of c urcumin and its implications for curcumin as a multipotent agent to combat Alzheimer's disease. *J. Mol. Struct. Theochem.* **2005**, *757*, 199–202. [CrossRef]
14. Nazari-Vanani, R.; Moezi, L.; Heli, H. In vivo evaluation of a self-nanoemulsifying drug delivery system for curcumin. *Biomed. Pharmacother.* **2017**, *88*, 715–720. [CrossRef]
15. Smirnova, E.; Moniruzzaman, M.; Chin, S.; Sureshbabu, A.; Karthikeyan, A.; Do, K.; Min, T. A Review of the Role of Curcumin in Metal Induced Toxicity. *Antioxidants* **2023**, *12*, 243. [CrossRef] [PubMed]
16. Prasad, S.; DuBourdieu, D.; Srivastava, A.; Kumar, P.; Lall, R. Metal–Curcumin Complexes in Therapeutics: An Approach to Enhance Pharmacological Effects of Curcumin. *Int. J. Mol. Sci.* **2021**, *22*, 7094. [CrossRef]
17. Vasdev, N.; Handa, M.; Kesharwani, P.; Shukla, R. Rosemary oil low energy nanoemulsion: Optimization, µrheology, in silico, in vitro, and ex vivo characterization. *J. Biomater. Sci. Polym. Ed.* **2022**, *33*, 1901–1923. [CrossRef] [PubMed]
18. Priyadarsini, K.I. The Chemistry of Curcumin: From Extraction to Therapeutic Agent. *Molecules* **2014**, *19*, 20091–20112. [CrossRef] [PubMed]
19. Hou, H.; Yu, Z.; Cai, S.; Nouman Shaukat, M.; Nazir, A.; Fallico, B. Ginger Bioactives: A Comprehensive Review of Health Benefits and Potential Food Applications. *Antioxidants* **2023**, *12*, 2015. [CrossRef]
20. Sarawi, W.S.; Alhusaini, A.M.; Fadda, L.M.; Alomar, H.A.; Albaker, A.B.; Aljrboa, A.S.; Alotaibi, A.M.; Hasan, I.H.; Mahmoud, A.M. Curcumin and nano-curcumin mitigate copper neurotoxicity by modulating oxidative stress, inflammation, and akt/gsk-3β signaling. *Molecules* **2021**, *26*, 5591. [CrossRef]
21. Kumar, R.; Uppal, S.; Kaur, K.; Mehta, S.K. Curcumin nanoemulsion as a biocompatible medium to study the metal ion imbalance in a biological system. *J. Mol. Liq.* **2020**, *314*, 113611. [CrossRef]
22. Handa, M.; Ujjwal, R.R.; Vasdev, N.; Flora, S.J.S.; Shukla, R. Optimization of Surfactant- And Cosurfactant-Aided Pine Oil Nanoemulsions by Isothermal Low-Energy Methods for Anticholinesterase Activity. *ACS Omega* **2021**, *6*, 559–568. [CrossRef]
23. Ningsih, I.Y.; Faradisa, H.; Cahyani, M.D.; Rosyidi, V.A.; Hidayat, M.A. The formulation of ginger oil nanoemulsions of three varieties of ginger (*Zingiber officinale* rosc.) as natural antioxidant. *J. Res. Pharm.* **2020**, *24*, 914–924. [CrossRef]
24. Elamin, E.; Jonkers, D.; Juuti-Uusitalo, K.; van IJzendoorn, S.; Troost, F.; Duimel, H.; Broers, J.; Verheyen, F.; Dekker, J.; Masclee, A. Effects of ethanol and acetaldehyde on tight junction integrity: In vitro study in a three dimensional intestinal epithelial cell culture model. *PLoS ONE* **2012**, *7*, e35008. [CrossRef]
25. Mhaske, A.; Kaur, J.; Naqvi, S.; Shukla, R. Decitabine enclosed biotin-zein conjugated nanoparticles: Synthesis, characterization, in vitro and in vivo evaluation. *Nanomedicine* **2024**, *19*, 1743–1760. [CrossRef] [PubMed]
26. Lewandowski, Ł.; Kepinska, M.; Milnerowicz, H. The copper-zinc superoxide dismutase activity in selected diseases. *Eur. J. Clin. Investig.* **2019**, *49*, e13036. [CrossRef] [PubMed]

27. Quamar, S.; Kumar, J.; Mishra, A.; Flora, S. Oxidative stress and neurobehavioural changes in rats following copper exposure and their response to MiADMSA and d-penicillamine. *Toxicol. Res. Appl.* **2019**, *3*, 239784731984478. [CrossRef]
28. Hirashima, Y.; Seshimo, S.; Fujiki, Y.; Okabe, M.; Nishiyama, K.; Matsumoto, M.; Kanouchi, H.; Oka, T. Homocysteine and copper induce cellular apoptosis via caspase activation and nuclear translocation of apoptosis-inducing factor in neuronal cell line SH-SY5Y. *Neurosci. Res.* **2010**, *67*, 300–306. [CrossRef]
29. Kumar, V.; Kalita, J.; Bora, H.K.; Misra, U.K. Temporal kinetics of organ damage in copper toxicity: A histopathological correlation in rat model. *Regul. Toxicol. Pharmacol.* **2016**, *81*, 372–380. [CrossRef] [PubMed]
30. Harwansh, R.K.; Mukherjee, P.K.; Bahadur, S.; Biswas, R. Enhanced permeability of ferulic acid loaded nanoemulsion based gel through skin against UVA mediated oxidative stress. *Life Sci.* **2015**, *141*, 202–211. [CrossRef]
31. Ahmed, K.; Li, Y.; McClements, D.J.; Xiao, H. Nanoemulsion- and emulsion-based delivery systems for curcumin: Encapsulation and release properties. *Food Chem.* **2012**, *132*, 799–807. [CrossRef]
32. Shafiq, S.; Shakeel, F.; Talegaonkar, S.; Ahmad, F.J.; Khar, R.K.; Ali, M. Development and bioavailability assessment of ramipril nanoemulsion formulation. *Eur. J. Pharm. Biopharm.* **2007**, *66*, 227–243. [CrossRef] [PubMed]
33. Jaiswal, M.; Dudhe, R.; Sharma, P.K. Nanoemulsion: An advanced mode of drug delivery system. *3 Biotech* **2015**, *5*, 123–127. [CrossRef]
34. Mahtab, A.; Anwar, M.; Mallick, N.; Naz, Z.; Jain, G.K.; Ahmad, F.J. Transungual Delivery of Ketoconazole Nanoemulgel for the Effective Management of Onychomycosis. *AAPS PharmSciTech* **2016**, *17*, 1477–1490. [CrossRef]
35. Mamale, K.; Shukla, S.; Mahale, P.; Mhaske, A.; Kaundal, R.K.; Shukla, R. Investigating the efficacy of gliclazide encapsulated hydrogel in the preclinical mice model for atopic dermatitis. In *Naunyn-Schmiedeberg's Archives of Pharmacology*; Springer: Berlin/Heidelberg, Germany, 2025; pp. 1–17.
36. Md, S.; Alhakamy, N.A.; Aldawsari, H.M.; Asfour, H.Z. Neuroprotective and Antioxidant Effect of Naringenin-Loaded Nanoparticles for Nose-to-Brain Delivery. *Brain Sci.* **2019**, *9*, 275. [CrossRef]
37. Giradkar, V.; Mhaske, A.; Shukla, R. Naringenin Nanocrystals Mitigate Rotenone Neurotoxicity in SH-SY5Y Cell Line by Modulating Mitophagy and Oxidative Stress. *AAPS PharmSciTech* **2024**, *25*, 227. [CrossRef]
38. Seo, S.W.; Han, H.K.; Chun, M.K.; Choi, H.K. Preparation and pharmacokinetic evaluation of curcumin solid dispersion using Solutol® HS15 as a carrier. *Int. J. Pharm.* **2012**, *424*, 18–25. [CrossRef]
39. Wang, Y.J.; Pan, M.H.; Cheng, A.L.; Lin, L.I.; Ho, Y.S.; Hsieh, C.Y.; Lin, J.K. Stability of curcumin in buffer solutions and characterization of its degradation products. *J. Pharm. Biomed. Anal.* **1997**, *15*, 1867–1876. [CrossRef]
40. Luo, Z.; Liu, Y.; Zhao, B.; Tang, M.; Dong, H.; Zhang, L.; Lv, B.; Wei, L. Ex vivo and in situ approaches used to study intestinal absorption. *J. Pharmacol. Toxicol. Methods* **2013**, *68*, 208–216. [CrossRef]
41. Othman, M.Z.; Hassan, Z.; Che Has, A.T. Morris water maze: A versatile and pertinent tool for assessing spatial learning and memory. *Exp. Anim.* **2022**, *71*, 264–280. [CrossRef]
42. Antunes, M.; Biala, G. The novel object recognition memory: Neurobiology, test procedure, and its modifications. *Cogn. Process.* **2011**, *13*, 93–110. [CrossRef] [PubMed] [PubMed Central]
43. Elwell, C.E.; Gagnon, N.L.; Neisen, B.D.; Dhar, D.; Spaeth, A.D.; Yee, G.M.; Tolman, W.B. Copper-Oxygen Complexes Revisited: Structures, Spectroscopy, and Reactivity. *Chem. Rev.* **2017**, *117*, 2059–2107. [CrossRef]

toxics

MDPI

Article

Does Personality Modulate the Sensitivity to Contaminants? A Case Study with Cadmium and Caffeine

Niedja Santos *, Sara Reis, Inês Domingues and Miguel Oliveira

Department of Biology & CESAM, University of Aveiro, Campus Universitario de Santiago, 3810-193 Aveiro, Portugal
* Correspondence: santos.niedja@ua.pt

Abstract: Personality has been reported to influence fish response to stress. This study aimed to assess whether shy and bold fish display different sensitivities to two environmental contaminants: caffeine (CAF) and cadmium (Cd). Thus, the sensitivity to Cd was compared based on lethal concentrations (LCs). The potential different response to CAF, known to alter the social behavior and locomotor activity of zebrafish, was studied using behavioral parameters. Overall, different LC values were found for each group: 48 h LC_{50} values of 4.79 (shy fish) and 8.20 $mg \cdot L^{-1}$ (bold fish); and 96 h LC_{50} values of 3.79 (shy fish) and 9.79 $mg \cdot L^{-1}$ (bold fish). In terms of response to CAF, a significant interaction between CAF and personality traits (bold and shy) was found in the locomotion activities (distance travelled, and medium and rapid movements), in the mirror test (frequency of contact and entries into the contact, approach, and distant zones), and in social tests (swimming distance in zones 2 and 3; time spent in zones 1, 2, and 3; and number of entries into zones 1 and 2). Shy fish exposed to 300 $\mu g \cdot L^{-1}$ of CAF presented hypoactivity, reduced aggressive behavior, and reduced sociability. Conversely, CAF did not influence the behavior of bold fish. In general, shy fish were more sensitive to Cd and exhibited anxious behavior when exposed to CAF, which appears to be the factor responsible for changes in their social behavior. Our results highlight the importance of taking personality traits into account in future studies, as variations in behavioral responses between bold and shy individuals can mask the toxicological effects of different chemicals.

Keywords: stress-coping styles; swimming behavior; short-term effects

1. Introduction

Personality traits are defined by individual behavioral traits that are persistent over time and that are expressed in different situations [1–5]. Cattell and Cattell [6] through empirical observation and statistical analysis, identified and defined 16 personality traits, including the bold and timid (shy) axis. The bold and shy axis stands out as a distinct, hereditary, and stable source of behavioral diversity [7]. Although this concept was initially described to detect individual differences in humans, personality traits are common among animals in the natural world [6,8]. In this regard, based on stable characteristics, including stress response, aggression, sociability, and the ability to explore a new environment, human and non-human individuals can be classified by personality [7,9,10]. Bolder profiles typically represent proactive individuals, dominant in social hierarchies, more aggressive, and more willing to take risks, and are also characterized by lower stress levels and a stronger inclination towards routines compared to their shy counterparts [7,9–11]. These inherent individual distinctions give rise to a range of cognitive skills in response to specific challenges, such as food and mate search, competition, and predator avoidance in fish [12]. Overall, bold and shy individuals display distinct susceptibilities to stressors. This divergence results from variations in how these individuals adapt to different environmental conditions, a process influenced by physiological differences (e.g., metabolic rate and neuroendocrine variations) [13,14].

The study of the links between personality traits and responses to various stressors in fish has been attracting the interest of the scientific community given the potential of personality traits to modulate the impacts of stressors on different organisms (e.g., fish) and consequently interfere in risk assessment. It has been reported that the effects of alcohol on shoaling, swimming speed, and exploration are modulated by personality traits [15–18]. Different responses have also been reported for nicotine (e.g., shy fish exhibited stronger anxiety-type responses than bold fish, as they reduced exploratory behavior [15]) and microplastics (bold zebrafish exhibited higher feeding activity levels, captured microplastics more frequently, and ingested a greater quantity of microplastics compared to shy zebrafish [19]).

Zebrafish have emerged as a good model for studying complex behaviors, including characteristics related to personality traits [15–23]. Different studies on fish personality usually assess the exploratory capacity (e.g., novel tank, T-labyrinth test, emergence test, feeding test), sociability (social test and/or shoaling test, mirror test), habituation to light–dark stimulus, and physiology of animals [2,24–26]. Nevertheless, there is a shortage of research regarding the shy and bold axis and its impact on responses to chemicals (e.g., pharmaceutical products) detected in the aquatic environment, as well as the resulting ecological consequences.

This study aimed to assess the sensitivity of personality traits (bold versus shy) to two chemicals known as environmental contaminants: cadmium (Cd) and caffeine (CAF). Human exposure to Cd and CAF has been associated with toxic effects. Cd toxicity is largely driven by oxidative stress, leading to DNA damage, impaired cell function, and organ toxicity in the heart, brain, liver, lungs, and kidneys [27]. Additionally, adverse effects of CAF consumption in healthy populations, particularly at high doses (above 400 mg/day), have been linked to cardiovascular, behavioural, reproductive, developmental, bone, and calcium disturbances [28,29].

Cd is a widely studied environmental pollutant and is considered a public health problem due to its widespread presence in various environmental compartments and its ability to cause toxicity in animals even at low concentrations, as well as its ability to bioaccumulate [30,31]. CAF is known as a marker of anthropogenic contamination and responsible for inducing behavioural alterations in zebrafish at concentrations ranging from μg to $mg \cdot L^{-1}$ [32–36]. In this regard, this study determined the median lethal concentration (LC_{50}) of Cd for bold and shy fish after 96 h of exposure and evaluated the behavioural responses (locomotion and social behaviour) of bold and shy fish after a 7-day exposure to CAF. We hypothesize that personality modulates the response to different environmental stressors.

2. Materials and Methods

2.1. Zebrafish Culture

Zebrafish (*D. rerio*) (AB strain) were kept in a ZebTEC (Tecniplast, Buguggiate, Italy) recirculating system at the University of Aveiro's Biology Department (Aveiro, Portugal). Culture water was generated by reverse osmosis, which was then complemented with salt (Instant Ocean Synthetic Sea Salt, Spectrum Brands). The water temperature was maintained at 27.0 ± 1 °C, conductivity at 800 ± 50 μS/cm, pH at 7.5 ± 0.5, and dissolved oxygen at ≥95% saturation. The photoperiod cycle (light–dark) was 14:10 h. Adult fish were fed once daily with GEMMA Micro 500 artificial feed (Skretting®, Algeciras, Spain). Adult zebrafish at 1 year old were chosen for the experimental assays. All procedures involving the use of animals were performed in accordance with European Union guidelines and regulations (e.g., OECD 203) and were approved by the Animal Care and Use Committee of the University of Aveiro, Portugal.

2.2. Personality Determination: Emergence Test

The selection of animals' personality based on an emergence test was conducted following the study performed by Mackenzie et al. [37]. The emergence test has been widely used to characterize bold and shy zebrafish based on their willingness to take risks. In this sense, the test was executed to assess an animal's tendency to leave a safe area and explore a new, less safe area. Briefly, rectangular glass tanks (40.6 cm × 26 cm × 19.5 cm) were used for acclimatization (acclimatization tanks) and for the test itself (test tanks). The test tank was divided into two sections: a small section (15 × 26 cm) where the fish are initially placed and a big section (25.6 × 26 cm). The two parts were divided by a black wall with a hatch that can be opened and closed and that allows fish to move from the small section to the big one. The day before to separation procedure, a set of 45 fish were transferred for a period of 12 h to acclimatization tanks to encourage them to explore the environment [38]. After acclimatization, 9 individuals were randomly selected and introduced into the small compartment of the test tank, with the hatch closed, for 10 min (Step 1). After this period, the hatch was open, allowing the fish to swim freely and explore the other compartment (Step 2). The first 3 fish to cross within 10 min were considered bold [37] and were immediately removed from the test tank and placed in a new tank. Steps 1 and 2 were repeated with the remaining fish, and the 3 individuals passing to the new compartment were labelled as intermediate (and were removed from the experiment). Those that stayed in the small compartment were subjected to Steps 1 and 2 again, but Step 2 was extended to 30 min. Fish that did not move to the other division were classified as shy, while those that did were discarded.

2.3. Chemicals

Cadmium chloride hemi(pentahydrate) (CAS number 7790-78-5; 99% purity) was obtained from Sigma Aldrich (Madrid, Spain) and was used as a source of cadmium (Cd).

Analytical grade CAF (1,3,7-Trimethylxanthine; CAS number 58-08-2; 98% purity) was obtained from TCI Chemicals (Zwijndrecht, Belgium).

The stock solution (Cd—250 mg·L^{-1}; CAF—10 mg·L^{-1}) and test solutions were prepared in the zebrafish water system (Section 2.1).

2.4. Sensitivity of Bold and Shy Fish to Cd

In order to determine the median lethal concentration (LC_{50}), the zebrafish previously selected as bold or shy were exposed for 96 h to 6 concentrations of Cd (0, 1.80, 3.00, 4.81, 7.70, and 12.31 mg·L^{-1}) selected based on preliminary range-finding tests. On the day of exposure, bold (n = 60) and shy (n = 60) male fish were distributed into 12 tanks (bold = 6 tanks; shy = 6 tanks); each tank contained 10 fish (shy or bold) and 1 L of test solution (0, 1.80, 3.00, 4.81, 7.70, or 12.31 mg·L^{-1}). The assay was based on the OECD testing guideline 203 on the Fish Acute Toxicity Test [39]. The temperature, the photoperiod, and the physical–chemical parameters of the water were kept equal to those of the cultivation (Section 2.1). During the test period, the medium was not renewed, and fish were not fed according to OECD 203 [39]. Mortality was recorded every 24 h up to 96 h, to determine LC_{50} values for bold and shy fish.

2.5. Exposure to CAF

Bold and shy fish ($n = 12$; ratio of 3 males to 1 female) were exposed to 3 concentrations of CAF (0, 1.5, and 300 $\mu g \cdot L^{-1}$) in semi-static settings over 7 days. The experiment was conducted in 20 L tank containing 5 L of the test solution and 12 fish per concentration. The concentrations tested were selected based on previous studies that reported the presence of these concentrations in the aquatic environment (e.g., 1.5 $\mu g \cdot L^{-1}$) [40,41] and their ability to induce changes in exploratory and social behavior of zebrafish exposed after 7 days of exposure (to 0.5. 1.5, and 300 $\mu g \cdot L^{-1}$) [36]. The medium was renewed every 2 days [42]. The temperature, the photoperiod, and the physical–chemical parameters of the water remained similar to the cultivation conditions (Section 2.1). During the experiment, the fish were fed once a day with the artificial diet Gemma Micro 500 (Skretting®, Algeciras, Spain) corresponding to 2% of the weight of the fish in each tank.

2.6. Behavior Assays

All fish from each concentration were subjected to the different behavioral tests. Figure 1 illustrates the experimental design and the sequence in which the tests were conducted.

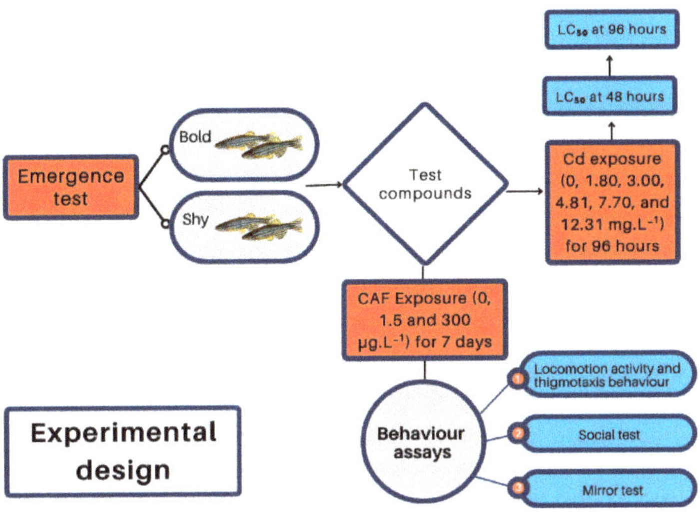

Figure 1. The flowchart illustrates the experimental process developed in this study. The first step involved conducting an emergence test to categorize the fish based on personality traits (bold and shy). The second step consisted of exposing the fish (bold and shy) to two chemicals: (1) Cd, to determine the median lethal concentration (LC_{50}) at 48 and 96 h of exposure; and (2) CAF, to assess behavioral changes. Behavioral tests were conducted in the following sequence: locomotor activity and thigmotaxis behavior, followed by the social test, and finally the mirror test.

2.6.1. Locomotor Activity and Thigmotaxis Behavior

Locomotion activity and thigmotaxis behavior were performed to assess anxiety-like behavior [43]. The fish were relocated individually into a new aquarium (9.4 cm wide and 14.1 cm long), and locomotor activity and thigmotaxis behavior were recorded using the ZebraBox™ (Viewpoint, Lyon, France) video tracking system. The test started with an acclimatization period (2 min) in the dark, followed by 2 min of light. The following parameters were examined: total distance (mm), swimming time (minutes), swimming speed in each area (slow—less than 8 mm/s; medium—between 8 and 40 mm/s; and rapid—greater than 40 mm/s), and fish path angles (class 1—angles from 180° to 90° and −180° to −90°; class 2—angles from −30° to −90° and from 30° to 90°; class 3—angles from −30° to −10° and 10° to 30°; and class 4—angles from −10° to 0° and 0° to 10°). To assess the fish's tendency to swim around the edges of the tank, (thigmotaxis behavior), the tank was virtually divided into two zones: an inner zone and an outer zone [43,44].

2.6.2. Mirror Biting Test

To study zebrafish aggressive behavior or sociability, a mirror image stimulation was used [45], following the protocol described by Santos et al. [36]. After experimental exposure, each fish was individually transferred to a rectangular tank (9 cm wide and 14 cm long) containing a mirror on one side. Fish behavior was recorded using the ZebraBox™ (Viewpoint, Lyon, France) video tracking system for 6 min, at 25 frames per second. The protocol consisted of an acclimatization period—1 min of darkness, followed by 5 min of light. For the behavior analysis, the tank was virtually divided in 3 zones: zone 1—zone of contact with the mirror (0.5 cm); zone 2—zone of approach (2.5 cm); and zone 3—zone far from the mirror (10 cm) [46]. The following parameters were analyzed: number of entries per zone, time spent in each zone, and swimming distance in each zone. The time spent in zone 1 was considered as the contact duration between the fish and the mirror. The videos recorded during the test were later analyzed for manual counting of the number of mirror contacts.

2.6.3. Social Test

Zebrafish are social animals that naturally form schools [47]. However, exposure to some chemicals modulates this behavior, increasing (e.g., 17-ethyl estradiol or carbaryl) or decreasing (e.g., CAF) school preference [36,43,48]. The methodology used in this study was adapted from the original protocol by Calcagno et al. [49], for the ZebraBox™ (Viewpoint, Lyon, France) video tracking system. The apparatus consisted of a tank (14 cm long × 6 cm high × 9 cm wide) with two compartments of different sizes (small, 6 cm long × 6 cm high × 9 cm wide; large, 8 cm long × 6 cm high × 9 cm wide) divided by a transparent barrier. The smaller compartment of the tank was used to house three medium-sized adult zebrafish that represented a school of zebrafish. These were placed in the tank before the test began. At the beginning of the experiment, one fish was introduced in the tank's largest compartment and then its behavior was recorded using the ZebraBox™ (Viewpoint, Lyon, France) video tracking system, for 6 min, at 25 frames per second. The protocol consisted of acclimatization period—1 min of darkness, followed by 5 min of light. For the analysis, the large compartment was virtually divided into three zones of equal size (approximately 2.6 cm): zone 1—zone close to the school; zone 2—intermediate zone; and zone 3—zone far from the school. Then, the swimming distance, swimming time, and number of entries into each zone were analyzed.

2.7. Statistical Analysis

Statistical analysis in this study was performed using SigmaPlot V.12.5 (SysStat Inc. software, Chicago, IL, USA). The medium lethal concentration (LC_{50}) of Cd was calculated using a 4-parameter logistic model (SigmaPlot 12.5 statistical package). The model choice was decided based on the R2 and the estimated residual standard error.

Data from locomotion, mirror, and social testing were analyzed using a one-way ANOVA (or Kruskal–Wallis test for non-normal distribution data), followed by multiple comparison testing. A two-way ANOVA (personality traits vs. concentration as factors) followed by a Holm–Sidak post-hoc test was performed to study the interaction between personality traits (bold and shy) and CAF. The level of significance for all statistical analyses was 0.05.

3. Results

3.1. Effects of Cd on Bold and Shy Zebrafish

The effects of Cd on bold and shy zebrafish survival were studied after 48 and 96 h of exposure (Figure 2). All shy zebrafish exposed to 7.70 $mg \cdot L^{-1}$ were dead after 24 h, whereas in bold zebrafish total mortality at 24 h was only found for 12.31 $mg \cdot L^{-1}$. The 48 h Cd LC_{50} for bold fish was 8.20 ± 0.035 $mg \cdot L^{-1}$, whereas for shy fish it was 4.79 ± 0.001 $mg \cdot L^{-1}$ (Figure 2A). After 96 h, the estimated LC_{50} of bold fish was 9.79 ± 2.372 $mg \cdot L^{-1}$, whereas for shy fish it was 3.79 ± 0.0447 $mg \cdot L^{-1}$ (Figure 2B).

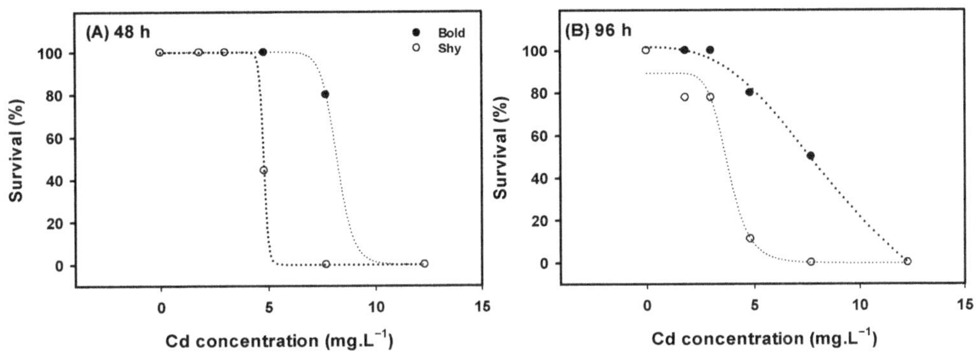

Figure 2. Survival curves of bold and shy zebrafish after exposure to cadmium (Cd) (n = 10). (**A**) Survival curve of bold and shy fish after exposure to Cd for 48 h, (**B**) survival curve of bold and shy fish after exposure to Cd for 96 h.

3.2. Effects of CAF on Bold and Shy Zebrafish

3.2.1. Locomotor Activity and Thigmotactic in Zebrafish

No significant differences between control bold and shy fish were found in terms of the distance travelled (Figure 3A), the percentage of distance travelled in slow movements (Figure S1, Supplementary Materials), or the distance moved in the edges of the tank (Figure 3D). However, control shy fish swam a significantly greater distance in medium-speed movements than the control bold fish (Figure 3B), whereas control bold fish swam greater distances in rapid movements (Figure 3C). No differences in swimming angles were found between the bold and shy control fish (Figure S1, Supplementary Materials).

Exposure to CAF did not induce significant effects on the locomotion of bold fish. However, shy fish upon exposure to the highest CAF concentration (300 $\mu g \cdot L^{-1}$) reduced the total distance travelled (ANOVA on ranks (Kruskall–Wallis) followed the Dunn's post-hoc test (H = 7.650, p = 0.022) (Figure 3A).

A two-way ANOVA followed by a Holm–Sidak post-hoc test revealed a significant interaction between CAF and personality traits in the distance travelled (p = 0.039), with shy fish decreasing their swimming activity at 300 $\mu g \cdot L^{-1}$. Bold fish showed no alterations in this parameter. A significant interaction between CAF and personality traits was also found in medium ($F_{(1,59)}$ = 14.861, $p < 0.001$) and rapid movements ($F_{(1,59)}$ = 14.861, $p < 0.001$), with shy fish exposed to 1.5 $\mu g \cdot L^{-1}$ CAF exhibiting an increase in the percentage of medium movements and a decrease in rapid movements while bold fish did not show alterations in these movements.

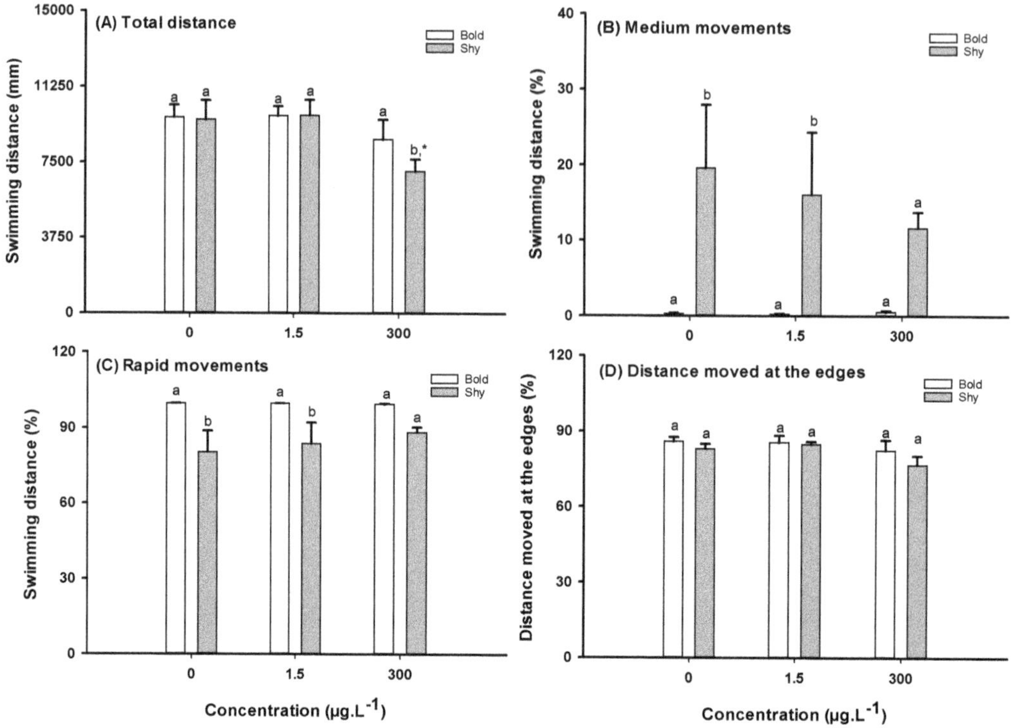

Figure 3. Effects on locomotion activity in bold and shy fish after 7 days of exposure to caffeine (CAF) (n = 12). (**A**) Total swimming distance; (**B**) distance travelled in medium movements, and (**C**) distance travelled in rapid movements; (**D**) percentage of distance moved at the edges of the tank by bold (white bars) and shy (grey bar) zebrafish. Data are presented as means ± standard errors. Different letters in bars indicate significant differences between bold and shy fish ($p < 0.05$). Asterisks (*) indicate differences from the respective control ($p < 0.05$).

3.2.2. Mirror Biting Test

No significant differences between bold and shy control fish were found in terms of the latency time to the first mirror approximation (Figure 4A), frequency of mirror bites (Figure 4B), or contact duration (Figure 4C).

Exposure to CAF significantly reduced the frequency of mirror biting in shy fish exposed to 300 µg·L^{-1} (one-way ANOVA followed by Dunnett's post-hoc test, $F_{(2,53)}$ = 0.05, p = 0.012) (Figure 4B) when compared to shy fish in the control group.

The influence of personality on the mirror biting test on the CAF was evaluated by performing a two-way ANOVA followed by a Holm–Sidak post-hoc test. A significant interaction between personality and CAF was observed in the frequency of mirror contact ($F_{(1,53)}$ = 14.861, p = 0.001), with shy fish showing a reduction in the frequency of contact with the mirror while bold fish did not exhibit any changes to this parameter. A significant interaction was also observed in the number of entries into zones 1 ($F_{(1,54)}$ = 6.222, p = 0.016), 2 ($F_{(1,54)}$ = 6.849, p = 0.011), and 3 ($F_{(1,53)}$ = 5.754, p = 0.020) of the tanks, with shy fish showing a decrease in the number of entries into zone 1 and an increase in the number of entries into zones 2 and 3 of the tank, while no changes were observed in bold fish.

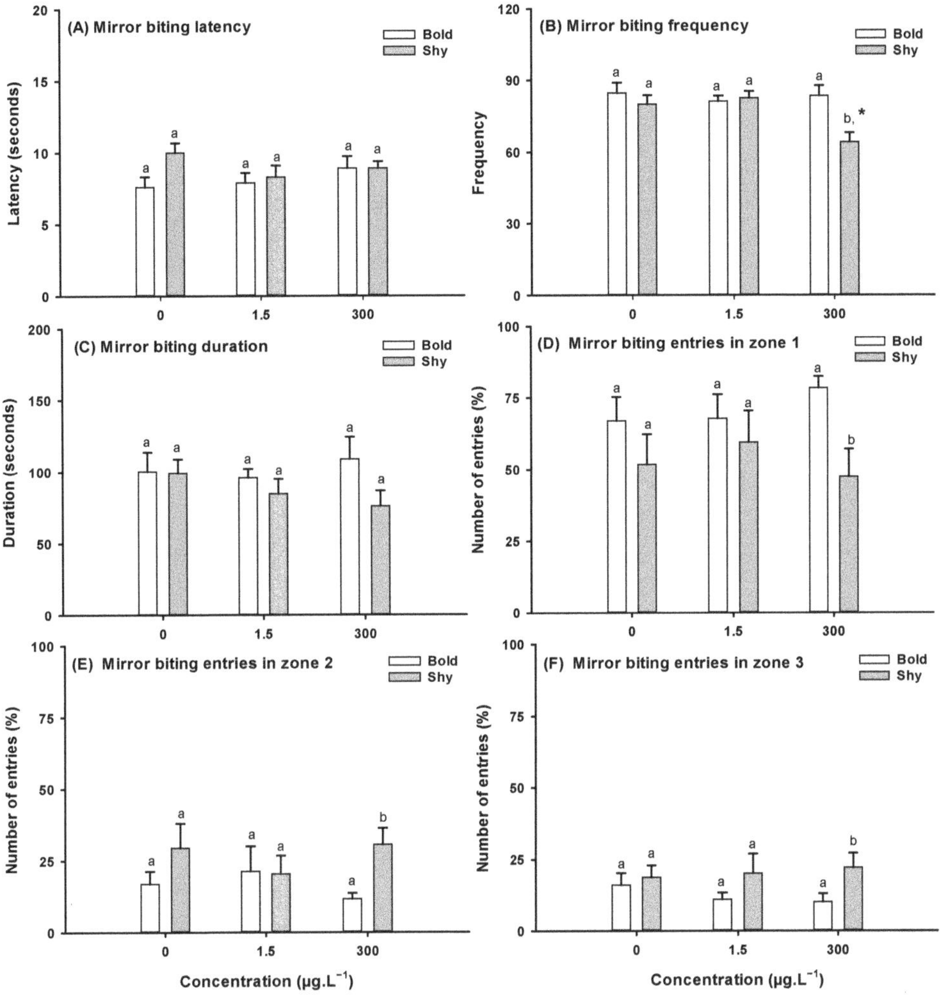

Figure 4. Effects on aggressive behavior in bold and shy fish after 7 days of exposure to caffeine (CAF) ($n = 12$). (**A**) Latency for the first contact with the mirror; (**B**) frequency of contact; (**C**) duration of contact; (**D**) number of entries into zone 1 (contact zone); (**E**) number of entries into zone 2 (approach zone); (**F**) number of entries into zone 3 (zone far from the mirror). Data are presented as means ± standard errors. Different letters in bars indicate significant differences between bold and shy fish. Asterisks (*) indicate differences from the respective control ($p < 0.05$).

3.2.3. Social Test

The social test revealed that bold and shy fish naturally exhibit distinct social behavior. According to a two-way ANOVA followed by a Holm–Sidak post-hoc test, bold control fish travelled a significantly greater distance in zone 3 ($F_{(1,58)} = 14.512$, $p < 0.012$) and spent more time in zone 1 ($F_{(1,59)} = 27.841$, $p = 0.004$) than shy control fish (Figure 5C,D). Shy control fish, on the other hand, spent significantly more time in zone 2 ($F_{(1,57)} = 14.384$, $p = 0.006$) (Figure 5E). Bold control fish also entered zone 3 ($F_{(1,59)} = 5.878$, $p = 0.031$) more frequently than shy control fish (Figure 5I).

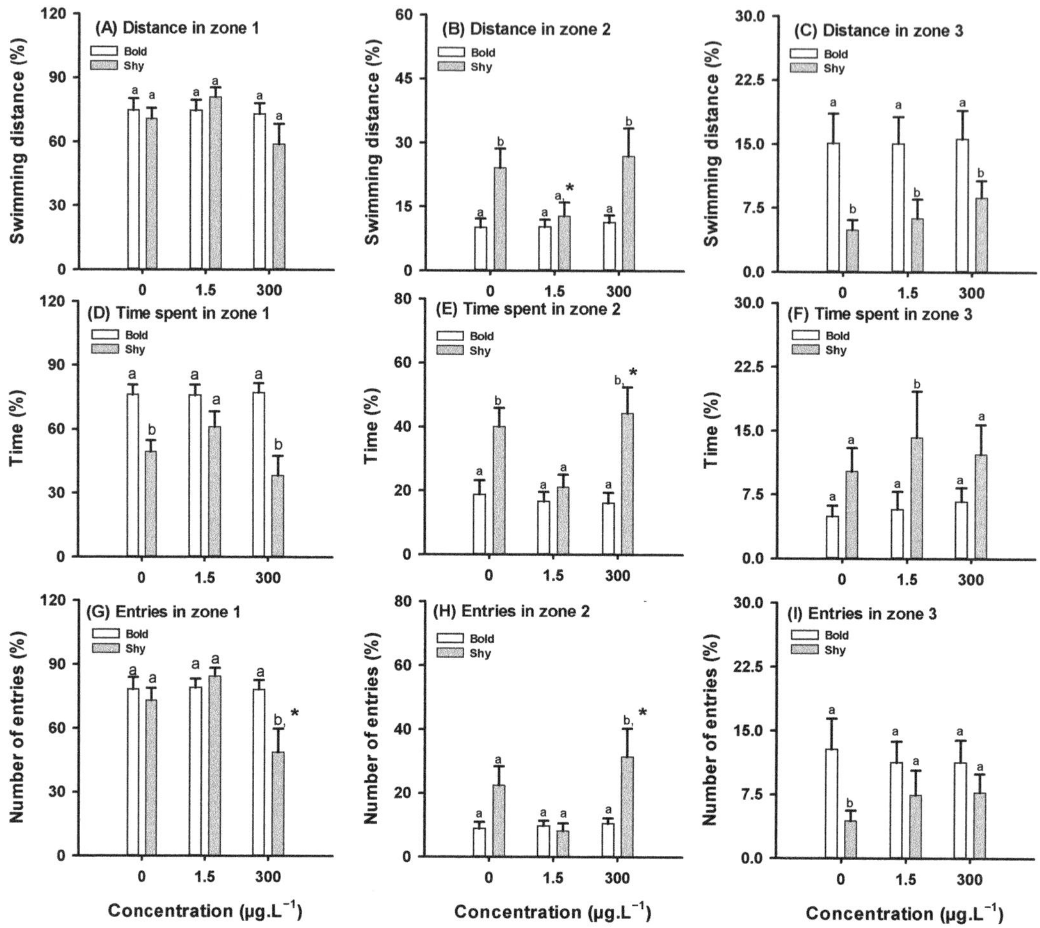

Figure 5. Effects on social behavior in bold and shy fish after 7 days of exposure to caffeine (CAF) (n = 12). (**A**) Swimming distance in zone 1; (**B**) swimming distance in zone 2; (**C**) swimming distance in zone 3; (**D**) time spent in zone 1; (**E**) time spent in zone 2; (**F**) time spent in zone 3; (**G**) number of entries into zone 1; (**H**) number of entries into zone 2; (**I**) number of entries into zone 3. Data are presented as means ± standard errors. Different letters in bars indicate significant differences between bold and shy fish ($p < 0.05$). Asterisks (*) indicate differences from the respective control ($p < 0.05$).

Exposure to CAF did not elicit significant effects on the social behavior of bold fish. However, shy fish increased or reduced social behaviors compared to the shy control depending on CAF concentration. CAF-exposed shy fish displayed significantly different behavior than their control organisms in a variety of parameters (e.g., distance travelled in zone 2, time spent in zone 2, and number of entries into zones 1 and 2). In terms of distance travelled (Figure 5B) and time spent in zone 2 (Figure 5E), fish exposed to CAF significantly lowered the distance travelled (1.5 μg·L^{-1}) ($F_{(2,57)} = 2.622$, $p = 0.048$) and increased the time spent (300 μg·L^{-1}) ($F_{(2,57)} = 3.809$, $p < 0.001$) when compared to the shy control group. Additionally, shy fish exposed to 300 μg·L^{-1} CAF significantly decreased the number of entrances into zone 1 ($F_{(2,57)} = 4.357$, $p = 0.019$), and increased in zone 2 ($F_{(2,57)} = 6.056$, $p < 0.001$) of the tank, when compared to the shy control group.

According to a two-way ANOVA followed by a Holm–Sidak post-hoc test, there was no interaction between the CAF and personality traits in terms of the distance travelled in zone 1 (Figure 5A) or the number of entries into zone 3 (Figure 5I). Interaction between CAF and personality traits was observed in the distance travelled in zone 2 ($F_{(1,57)} = 14.384$, $p < 0.001$) (Figure 5C) and zone 3 ($F_{(1,57)} = 14.512$, $p < 0.001$) (Figure 5B) of the tank, with shy fish exposed to CAF showing an increase in the distance travelled in zone 2 and in zone 3, while bold fish did not exhibit changes in the distance travelled in these two areas. A significant interaction was observed in the time travelled in zones 1 ($F_{(1,59)} = 27.841$, $p < 0.001$) (Figure 5D), 2 ($F_{(1,59)} = 20.863$, $p < 0.001$) (Figure 5F), and 3 ($F_{(1,59)} = 6.682$, $p = 0.012$) (Figure 5E) of the tank, with shy fish showing a decrease in time spent in zone 1 and an increase in time spent in zone 2 and zone 3, while bold fish did not show any change in the time spent in any of these areas (Figure 5D–F). A significant interaction was also observed in the number of entries into zone 1 ($F_{(1,59)} = 3.565$, $p = 0.002$) (Figure 5G) and 2 ($F_{(1,59)} = 12.129$, $p < 0.001$) (Figure 5H) of the tank, with shy fish showing a decrease in the number of entries into zone 1, and an increase in the number of entries into zone 2, while fish bold did not significantly change the number of entries into different areas of the tank (Figure 5G–I).

4. Discussion

This study analyzed the sensitivity of personality traits (bold and shy) to cadmium (Cd) and caffeine (CAF). In the case of Cd, clear differences in Cd sensitivity between bold and shy fish were observed. At 48 and 96 h of exposure, bold fish exhibited greater resistance to Cd as evidenced by a higher LC_{50} value compared to shy fish (Figure 2A,B). Cd and related compounds are environmental pollutants, known for their capacity to cause toxicity in animals and humans, as well as for their ability to bioaccumulate [30,31,50]. LC_{50} values for Cd were previously determined in zebrafish (disregarding personality traits). The estimated LC_{50} values for the same periods tested in the present study were 12.88 $mg \cdot L^{-1}$ at 48 h and 9.68 $mg \cdot L^{-1}$ at 96 h [31]. In the present study, the LC_{50} values of Cd for bold fish were 1.7 and 2.6 times higher than those for shy fish at 48 and 96 h, respectively. One of the pathways by which Cd exerts toxicity in fish is through the increased synthesis of reactive oxygen species (ROS) [51]. It has been observed that shy individuals exhibit higher levels of ROS and subsequent oxidative stress due to elevated stress hormone levels when compared to bold fish [52]. In this context, higher basal ROS levels in shy fish add to the ROS induced by Cd exposure, leading to the higher susceptibility of shy fish towards the compound. In this sense, differences in personality traits may serve as predictors of susceptibility to compounds and physical fitness [53]. However, further studies addressing ROS levels and oxidative stress enzymes activities in both bold and shy zebrafish exposed to Cd are needed for a clearer understanding.

Previous studies revealed that bold fish are more active than shy fish, explore new environments more, spend more time in unprotected areas, and have a greater speed while moving [16,20,54–56]. Our data (distance travelled by the control in slow movements, medium movements, and rapid movements) are in agreement with those studies. In terms of locomotion activity (total distance travelled; distance travelled in medium and rapid movements), an interaction between CAF and personality traits could be observed. For example, exposure to 1.5 $\mu g \cdot L^{-1}$ did not change the distance travelled by shy fish in slow, medium, and rapid movements when compared to the shy control fish. However, they displayed a higher percentage of medium movements and a lower percentage of rapid movements when compared to bold fish subjected to the same conditions (Figure 3B,C). Shy fish exposed to 300 $\mu g \cdot L^{-1}$ CAF presented hypoactivity (reduced distance travelled), while bold fish showed no differences in these parameters. Hypoactivity associated with increased anxiety levels after exposure to CAF is a behavior commonly described in both larvae and adult zebrafish [33,57]. Santos et al. [36] identified that a 7-day exposure of adult zebrafish to 300 $\mu g \cdot L^{-1}$ of CAF did not have a significant impact on their locomotion behavior (e.g., distance travelled), although it reduced the vertical exploratory behavior of these fish. In the locomotion test performed in the present study, CAF interacted differently with bold and shy fish. Broadly, bold fish did not display behavioral change when exposed to CAF compared to the control group, which suggests the low susceptibility of these fish to CAF. Exposed shy fish showed significant behavioral variations when compared to shy controls, indicating higher susceptibility to CAF. Factors like variations in metabolic rates between bold and shy fish or the presence of a greater number of type A_1 adenosine receptors, which are linked to decreased zebrafish swimming activity following CAF exposure, may account for differences in their locomotor activity [58,59]. Differences in bold and shy responses appear to vary depending on the chemical they are exposed to, e.g., in agreement with the effects of CAF observed in this study, such as the increased susceptibility of shy fish. Araujo-Silva et al. [15] reported that shy zebrafish displayed anxiety-like responses following a 60 min exposure to 1 $mg \cdot L^{-1}$ of nicotine when compared to bold fish. On the other hand, bold zebrafish are more susceptible to alcohol than shy fish, as they exhibit anxiety-like behavior when exposed to 0.1% alcohol for 60 min [15].

Bold and shy brook trout (*Salvelinus fontinalis*) show different levels of aggression [56]. In this study, short-term CAF (300 $\mu g \cdot L^{-1}$) exposure significantly reduced aggressive behavior (reduced the frequency of contact with the mirror and the number of entries into zone 1) or caused shy fish to become more cautious but did not change bold fish behavior. Exposure to CAF is linked to heightened aggression and anxiety in zebrafish. Zebrafish anxious behavior is linked to a decrease in exploratory behavior [33,36,57] and risk-taking. As demonstrated by the locomotion test (e.g., total distance) and the reduced mirror approach, shy fish exposed to 300 $\mu g \cdot L^{-1}$ of CAF show anxious behavior that is not observed in bold fish. In the mirror test, the shy responses to the mirror image stimulus are in accordance with what has previously been described, i.e., in the presence of acute stress, shy fish tend to respond passively, seeking to "hide/freeze" [60]. On the other hand, it was expected that bold fish would respond actively through "fight/flight" behavior [60]; however, fight behavior was not detected, which may suggest that the CAF concentrations tested were insufficient to alter the behavioral pattern in bold fish. A study carried out by Santos et al. [36] demonstrated that exposure to 0.5, 1.5, and 300 $\mu g \cdot L^{-1}$ of CAF for 7 days induced aggressive behavior in zebrafish (in undifferentiated fish). On the other hand, similar to the results observed in the present study with shy fish in the mirror test, Gutiérrez et al. [61] reported that a 30 min exposure to 19.4 $\mu g \cdot L^{-1}$ of CAF led to a reduction in mirror aggression in undifferentiated juvenile zebrafish. Differences in responses between studies (e.g., [36,61]) can be explained by the presence of personality traits. For example, in this study, bold fish exposed to CAF did not change their response to the mirror image; however, shy fish have reduced aggressive behavior. In general, anxiety influences social behaviors. In rats, anxiety-like behaviors encompass social avoidance [62]. The results from the social test performed in this study also support the hypothesis of increased anxiety observed in shy individuals exposed to 300 $\mu g \cdot L^{-1}$ of CAF, as suggested by locomotion activity and the mirror test results. Exposure to 300 $\mu g \cdot L^{-1}$ of CAF significantly reduced the sociability of shy fish. In zebrafish, a decrease in social interaction has also been linked to increased anxiety [63]. A study carried out by Santos et al. [36] showed that the exposure of undifferentiated zebrafish to CAF for 7 days reduced sociability at CAF concentrations of 0.5 and 1.5 $\mu g \cdot L^{-1}$. However, it did not affect fish social behavior when exposed to 300 $\mu g \cdot L^{-1}$ of CAF. Results similar to those found in this study on social behavior have been described by Araujo-Silva [17], who described that the exposure of shy fish to 0.1% of alcohol for 60 min reduced the sociability of fish. In this regard, the results observed in the present study in both shy and bold zebrafish during the social test agree with the results of the locomotion activity and the mirror test, confirming the difference in the consistent behavioral pattern between these different personality traits.

In general, it was observed that, in this study, the CAF concentrations tested were not able to alter the behavior of bold fish but induced behavioral changes in shy fish. It is known that individual fish behaviors are controlled by endogenous and exogenous factors [60]. Therefore, individual characteristics can explain the differences between the results described by Santos et al. [36] and the results of the present study. For example, the increase in anxiety levels associated with CAF consumption is linked to its interaction with A_1-like receptors [58]. The exposure of zebrafish to CAF from 1 h post fertilization can increase A_1 expression [64]. Therefore, variation in the gene expression of these receptors between bold and shy zebrafish could influence the anxiety level displayed by these fish when exposed to CAF. Furthermore, there is evidence that dopamine plays an important role in regulating anxiety. Thörnqvist et al. [55] found that bold zebrafish, characterized by higher risk-taking behavior and increased locomotor activity, exhibit elevated expression of D_2-type dopamine receptors (drd2a and drd2b) in comparison to shy fish. In general, there are several physiological variations between bold and shy fish, such as changes in metabolic rate and the generation of ROS, which can influence the rate of metabolization and chemical excretion [52,60,65]. In this sense, more studies are needed to understand the differences in CAF metabolism rates and the mechanisms involved in the responses of bold and shy individuals when exposed to chemicals.

5. Conclusions

This study evaluated the sensitivity of bold and shy zebrafish when exposed to cadmium (Cd) or caffeine (CAF), two well-studied environmental contaminants. There are differences in Cd sensitivity between bold and shy fish, with bold zebrafish being more resistant to Cd. Shy fish are also more sensitive to CAF. CAF induced anxious behavior in shy fish, detected by reduced locomotion activity, reduced aggressive behavior, and sociability, but had no effect on bold fish. Overall, our findings highlight those differences between shy and bold individuals can influence fish responses during chemical exposure. In conventional tests, where fish are not differentiated based on personality traits, the response to chemicals can vary depending on the proportion of bold and shy fish. The lack of classification of fish by personality traits may mask behavioral effects after chemical exposure. Studying personality traits may reduce the variability in behavioral responses described in the literature in chemically exposed fish.

Supplementary Materials: The following supporting information can be downloaded at: https://www.mdpi.com/article/10.3390/toxics13030147/s1, Figure S1: Behavioural effects on bold and shy fish after 7 days of exposure to Caffeine (CAF) (n = 12).

Author Contributions: N.S.: Conceptualization, methodology, and writing—original draft, review, and editing; S.R.: Methodology; I.D.: Conceptualization, validation, and writing—review and editing; M.O.: Conceptualization, validation, and writing—review and editing. All authors have read and agreed to the published version of the manuscript.

Funding: This work was developed within the project NanoPlanet (2022.02340.PTDC), financially supported by National Funds (OE), through FCT/MCTES. Thanks are also due for the financial support to CESAM financed by National Funds through FCT/MCTES (UIDB/50017/2020, UIDP/50017/2020, and LA/P/0094/2020) through strategic programs and the PhD grant awarded to Niedja Santos (BD/REIT/8708/2019).

Institutional Review Board Statement: All experimental procedures involving fish were carried out in compliance with the International Guiding Principles for Biomedical Research Involving Animals (EU Directive 2010/63) and the regulations of the Portuguese Directorate-General for Food and Veterinary Affairs (DGAV) on animal welfare. The study was approved by the Ethics Committee of the University of Aveiro and the DGAV on 5 June 2019 (protocol number 0421/000/000/000/2019).

Informed Consent Statement: Not applicable.

Data Availability Statement: The original data presented in the study are openly available in the Supplementary Materials.

Conflicts of Interest: The authors declare no conflicts of interest.

Abbreviations

The following abbreviations are used in this manuscript:

ANOVA	Analysis of variance
Cd	Cadmium
CAF	Caffeine
LC	Lethal concentration
LC_{50}	Median lethal concentration
OECD	The Organization for Economic Cooperation and Development
ROS	Reactive oxygen species

References

1. Alfonso, S.; Sadoul, B.; Gesto, M.; Joassard, L.; Chatain, B.; Geffroy, B.; Bégout, M.-L. Coping styles in European sea bass: The link between boldness, stress response and neurogenesis. *Physiol. Behav.* **2019**, *207*, 76–85. [CrossRef] [PubMed]
2. Colchen, T.; Faux, E.; Teletchea, F.; Pasquet, A. Is personality of young fish consistent through different behavioural tests? *Appl. Anim. Behav. Sci.* **2017**, *194*, 127–134. [CrossRef]
3. Koolhaas, J.M.; Korte, S.M.; De Boer, S.F.; Van Der Vegt, B.J.; Van Reenen, C.G.; Hopster, H.; De Jong, I.C.; Ruis, M.A.W.; Blokhuis, H.J. Coping styles in animals: Current status in behavior and stress-physiology. *Neurosci. Biobehav. Rev.* **1999**, *23*, 925–935. [CrossRef]
4. Neave, H.W.; Costa, J.H.C.; Weary, D.M.; von Keyserlingk, M.A.G. Long-term consistency of personality traits of cattle. *R. Soc. Open Sci.* **2020**, *7*, 191849. [CrossRef] [PubMed]
5. O'malley, C.I.; Turner, S.P.; D'eath, R.B.; Steibel, J.P.; Bates, R.O.; Ernst, C.W.; Siegford, J.M. Animal personality in the management and welfare of pigs. *Appl. Anim. Behav. Sci.* **2019**, *218*, 104821. [CrossRef]
6. Cattell, R.B.; Cattell, H.E.P. Personality Structure and the New Fifth Edition of the 16PF. *Educ. Psychol. Meas.* **1995**, *55*, 926–937. [CrossRef]
7. Toms, C.N.; Echevarria, D.J.; Jouandot, D.J. A methodological review of personality-related studies in fish: Focus on the shy-bold axis of behavior. *Int. J. Comp. Psychol.* **2010**, *23*, 1–25. [CrossRef]
8. Kelleher, S.R.; Silla, A.J.; Byrne, P.G. Animal personality and behavioral syndromes in amphibians: A review of the evidence, experimental approaches, and implications for conservation. *Behav. Ecol. Sociobiol.* **2018**, *72*, 79. [CrossRef]
9. Rey, S.; Digka, N.; MacKenzie, S. Animal personality relates to thermal preference in wild-type zebrafish, *Danio rerio*. *Zebrafish* **2015**, *12*, 243–249. [CrossRef] [PubMed]
10. Siviter, H.; Deeming, D.C.; Rosenberger, J.; Burman, O.H.P.; Moszuti, S.A.; Wilkinson, A. The impact of egg incubation temperature on the personality of oviparous reptiles. *Anim. Cogn.* **2017**, *20*, 109–116. [CrossRef] [PubMed]
11. Bensky, M.K.; Paitz, R.; Pereira, L.; Bell, A.M. Testing the predictions of coping styles theory in threespined sticklebacks. *Behav. Process.* **2017**, *136*, 1–10. [CrossRef]
12. Lucon-Xiccato, T.; Montalbano, G.; Bertolucci, C. Personality traits covary with individual differences in inhibitory abilities in 2 species of fish. *Curr. Zool.* **2019**, *66*, 187–195. [CrossRef]
13. Castanheira, M.F.; Conceição, L.E.; Millot, S.; Rey, S.; Bégout, M.; Damsgård, B.; Kristiansen, T.; Höglund, E.; Øverli, Ø.; Martins, C.I. Coping styles in farmed fish: Consequences for aquaculture. *Rev. Aquac.* **2017**, *9*, 23–41. [CrossRef]
14. Larsen, M.H.; Johnsson, J.I.; Winberg, S.; Wilson, A.D.M.; Hammenstig, D.; Thörnqvist, P.-O.; Midwood, J.D.; Aarestrup, K.; Höglund, E. Effects of emergence time and early social rearing environment on behaviour of Atlantic salmon: Consequences for juvenile fitness and smolt migration. *PLoS ONE* **2015**, *10*, e0119127. [CrossRef]

15. Araujo-Silva, H.; de Souza, A.M.; Mamede, J.P.M.; de Medeiros, S.R.B.; Luchiari, A.C. Individual differences in response to alcohol and nicotine in zebrafish: Gene expression and behavior. *Dev. Growth Differ.* **2023**, *65*, 434–445. [CrossRef] [PubMed]
16. Araujo-Silva, H.; Leite-Ferreira, M.E.; Luchiari, A.C. Behavioral screening of alcohol effects and individual differences in zebrafish (*Danio rerio*). *Alcohol Alcohol.* **2020**, *55*, 591–597. [CrossRef]
17. Araujo-Silva, H.; Pinheiro-da-Silva, J.; Silva, P.F.; Luchiari, A.C. Individual differences in response to alcohol exposure in zebrafish (*Danio rerio*). *PLoS ONE* **2018**, *13*, e0198856. [CrossRef] [PubMed]
18. Bellot, M.S.; Guermandi, I.I.; Camargo-dos-Santos, B.; Giaquinto, P.C. Differences in the alcohol preference assessment of shy and bold zebrafish. *Front. Behav. Neurosci.* **2022**, *16*, 810051. [CrossRef]
19. Chen, Y.; Li, W.; Xiang, L.; Mi, X.; Duan, M.; Wu, C. Fish personality affects their exposure to microplastics. *Ecotoxicol. Environ. Saf.* **2022**, *233*, 113301. [CrossRef] [PubMed]
20. Rajput, N.; Parikh, K.; Kenney, J.W. Beyond bold versus shy: Zebrafish exploratory behavior falls into several behavioral clusters and is influenced by strain and sex. *Biol. Open* **2022**, *11*, bio059443. [CrossRef] [PubMed]
21. Santos, C.P.; de Oliveira, M.N.; Silva, P.F.; Luchiari, A.C. Relationship between boldness and exploratory behavior in adult zebrafish. *Behav. Process.* **2023**, *209*, 104885. [CrossRef] [PubMed]
22. Daniel, D.K.; Bhat, A. Correlations begin at home: Drivers of co-occurrence patterns in personality and cognitive ability in wild populations of zebrafish. *Anim. Cogn.* **2023**, *26*, 1381–1394. [CrossRef] [PubMed]
23. Axling, J.; Jakobsson, H.; Frymus, N.; Thörnqvist, P.O.; Petersson, E.; Winberg, S. Boldness in zebrafish larvae—Development and differences between a domesticated lab strain and offspring of wild-caught fish. *Fishes* **2022**, *7*, 197. [CrossRef]
24. Castanheira, M.F.; Cerqueira, M.; Millot, S.; Gonçalves, R.A.; Oliveira, C.C.; Conceição, L.E.; Martins, C.I. Are personality traits consistent in fish?-The influence of social context. *Appl. Anim. Behav. Sci.* **2016**, *178*, 96–101. [CrossRef]
25. Ferrari, S.; Millot, S.; Leguay, D.; Chatain, B.; Bégout, M.L. Consistency in European seabass coping styles: A life-history approach. *Appl. Anim. Behav. Sci.* **2015**, *167*, 74–88. [CrossRef]
26. Réale, D.; Reader, S.M.; Sol, D.; McDougall, P.T.; Dingemanse, N.J. Integrating animal temperament within ecology and evolution. *Biol. Rev.* **2007**, *82*, 291–318. [CrossRef] [PubMed]
27. Vamsi, N.M.; Kumar, J.P.; Ramadevi, K.; Swathi, P. Cadmium toxicity: Unveiling the threat to human health. *Indian. J. Pharm. Sci.* **2024**, *86*, 1601–1610. [CrossRef]
28. Wikoff, D.; Welsh, B.T.; Henderson, R.; Brorby, G.P.; Britt, J.; Myers, E.; Goldberger, J.; Lieberman, H.R.; O'Brien, C.; Peck, J.; et al. Systematic review of the potential adverse effects of caffeine consumption in healthy adults, pregnant women, adolescents, and children. *Food Chem. Toxicol.* **2017**, *109*, 585–648. [CrossRef]
29. Nawrot, P.; Jordan, S.; Eastwood, J.; Rotstein, J.; Hugenholtz, A.; Feeley, M. Effects of caffeine on human health. *Food Addit. Contam.* **2003**, *20*, 1–30. [CrossRef]
30. Flick, D.F.; Kraybill, H.F.; Dlmitroff, J.M. Toxic effects of cadmium: A review. *Environ. Res.* **1971**, *4*, 71–85. [CrossRef]
31. Al-sawafi, A.G.A.; Wang, L.; Yan, Y. Cadmium accumulation and its histological effect on brain and skeletal muscle of zebrafish. *J. Heavy Met. Toxic. Dis.* **2017**, *2*. [CrossRef]
32. Santos, L.C.; Ruiz-Oliveira, J.; Oliveira, J.J.; Silva, P.F.; Luchiari, A.C. Irish coffee: Effects of alcohol and caffeine on object discrimination in zebrafish. *Pharmacol. Biochem. Behav.* **2016**, *143*, 34–43. [CrossRef] [PubMed]
33. Rosa, L.V.; Ardais, A.P.; Costa, F.V.; Fontana, B.D.; Quadros, V.A.; Porciúncula, L.O.; Rosemberg, D.B. Different effects of caffeine on behavioral neurophenotypes of two zebrafish populations. *Pharmacol. Biochem. Behav.* **2018**, *165*, 1–8. [CrossRef]
34. Tran, S.; Fulcher, N.; Nowicki, M.; Desai, P.; Tsang, B.; Facciol, A.; Chow, H.; Gerlai, R. Time-dependent interacting effects of caffeine, diazepam, and ethanol on zebrafish behaviour. *Prog. Neuropsychopharmacol. Biol. Psychiatry* **2017**, *75*, 16–27. [CrossRef] [PubMed]
35. Santos, L.C.; Ruiz-Oliveira, J.; Silva, P.F.; Luchiari, A.C. Caffeine dose-response relationship and behavioral screening in zebrafish. In *The Question of Caffeine*; InTech: London, UK, 2017. [CrossRef]
36. Santos, N.; Picolo, V.; Domingues, I.; Perillo, V.; Villacis, R.A.; Grisolia, C.K.; Oliveira, M. Effects of environmental concentrations of caffeine on adult zebrafish behaviour: A short-term exposure scenario. *Environ. Sci. Pollut. Res.* **2023**, *30*, 63776–63787. [CrossRef]
37. MacKenzie, S.; Ribas, L.; Pilarczyk, M.; Capdevila, D.M.; Kadri, S.; Huntingford, F.A. Screening for coping style increases the power of gene expression studies. *PLoS ONE* **2009**, *4*, e5314. [CrossRef] [PubMed]
38. Blaser, R.E.; Rosemberg, D.B. Measures of anxiety in zebrafish (*Danio rerio*): Dissociation of black/white preference and novel tank test. *PLoS ONE* **2012**, *7*, e36931. [CrossRef]
39. OECD. *Guidelines for the Testing of Chemicals: 203—Fish, Acute Toxicity Test*; OECD: Paris, France, 1992.
40. Nödler, K.; Voutsa, D.; Licha, T. Polar organic micropollutants in the coastal environment of different marine systems. *Mar. Pollut. Bull.* **2014**, *85*, 50–59. [CrossRef]
41. Siegener, R.; Chen, R.F. Caffeine in Boston Harbor seawater. *Mar. Pollut. Bull.* **2002**, *44*, 383–387. [CrossRef] [PubMed]

42. Lam, M.W.; Young, C.J.; Brain, R.A.; Johnson, D.J.; Hanson, M.A.; Wilson, C.J.; Richards, S.M.; Solomon, K.R.; Mabury, S.A. Aquatic persistence of eight pharmaceuticals in a microcosm study. *Environ. Toxicol. Chem.* **2004**, *23*, 1431–1440. [CrossRef] [PubMed]
43. Correia, D.; Almeida, A.R.; Santos, J.; Machado, A.L.; Ucun, O.K.; Žlábek, V.; Oliveira, M.; Domingues, I. Behavioral effects in adult zebrafish after developmental exposure to carbaryl. *Chemosphere* **2019**, *235*, 1022–1029. [CrossRef]
44. Santos, N.; Domingues, I.; Oliveira, M. The role of temperature on zebrafish ontogenic development and sensitivity to pharmaceuticals. *Environ. Toxicol. Pharmacol.* **2023**, *103*, 104256. [CrossRef] [PubMed]
45. Gerlai, R.; Lahav, M.; Guo, S.; Rosenthal, A. Drinks like a fish: Zebrafish (*Danio rerio*) as a behavior genetic model to study alcohol effects. *Pharmacol. Biochem. Behav.* **2000**, *67*, 773–782. [CrossRef]
46. Pham, M.; Raymond, J.; Hester, J.; Kyzar, E.; Gaikwad, S.; Bruce, I.; Fryar, C.; Chanin, S.; Enriquez, J.; Bagawandoss, S.; et al. Assessing social behavior phenotypes in adult zebrafish: Shoaling, social preference, and mirror biting tests. In *Zebrafish Protocols for Neurobehavioral Research*; Humana Press: Totowa, NJ, USA, 2012; pp. 231–246. [CrossRef]
47. Gerlai, R. Zebrafish antipredatory responses: A future for translational research? *Behav. Brain Res.* **2010**, *207*, 223–231. [CrossRef]
48. Reyhanian, N.; Volkova, K.; Hallgren, S.; Bollner, T.; Olsson, P.-E.; Olsén, H.; Hällström, I.P. 17α-Ethinyl estradiol affects anxiety and shoaling behavior in adult male zebra fish (*Danio rerio*). *Aquat. Toxicol.* **2011**, *105*, 41–48. [CrossRef]
49. Calcagno, E.; Durando, P.; Valdés, M.E.; Franchioni, L.; de los Ángeles Bistoni, M. Effects of carbamazepine on cortisol levels and behavioral responses to stress in the fish Jenynsia multidentata. *Physiol. Behav.* **2016**, *158*, 68–75. [CrossRef]
50. Taslima, K.; Emran, A.; Rahman, M.S.; Hasan, J.; Ferdous, Z.; Rohani, F. Impacts of heavy metals on early development, growth and reproduction of fish—A review. *Toxicol. Rep.* **2022**, *9*, 858–868. [CrossRef] [PubMed]
51. Lee, J.W.; Jo, A.H.; Lee, D.C.; Choi, C.Y.; Kang, J.C.; Kim, J.H. Review of cadmium toxicity effects on fish: Oxidative stress and immune responses. *Environ. Res.* **2023**, *236*, 116600. [CrossRef]
52. Arnold, K.E.; Herborn, K.A.; Adam, A.; Alexander, L. Individual variation in the oxidative costs of personality traits. *Funct. Ecol.* **2015**, *29*, 522–530. [CrossRef]
53. Herborn, K.A.; Coffey, J.; Larcombe, S.D.; Alexander, L.; Arnold, K.E. Oxidative profile varies with personality in European greenfinches. *J. Exp. Biol.* **2011**, *214*, 1732–1739. [CrossRef] [PubMed]
54. Sneddon, L.U. The bold and the shy: Individual differences in rainbow trout. *J. Fish. Biol.* **2003**, *62*, 971–975. [CrossRef]
55. PThörnqvist, O.; McCarrick, S.; Ericsson, M.; Roman, E.; Winberg, S. Bold zebrafish (*Danio rerio*) express higher levels of delta opioid and dopamine D2 receptors in the brain compared to shy fish. *Behav. Brain Res.* **2019**, *359*, 927–934. [CrossRef]
56. White, S.L.; Wagner, T.; Gowan, C.; Braithwaite, V.A. Can personality predict individual differences in brook trout spatial learning ability? *Behav. Process.* **2017**, *141 Pt 2*, 220–228. [CrossRef]
57. de Farias, N.O.; Andrade, T.d.S.; Santos, V.L.; Galvino, P.; Suares-Rocha, P.; Domingues, I.; Grisolia, C.K.; Oliveira, R. Neuromotor activity inhibition in zebrafish early-life stages after exposure to environmental relevant concentrations of caffeine. *J. Environ. Sci. Health Part A* **2021**, *56*, 1306–1315. [CrossRef] [PubMed]
58. El Yacoubi, M.; Ledent, C.; Ménard, J.F.; Parmentier, M.; Costentin, J.; Vaugeois, J.M. The stimulant effects of caffeine on locomotor behaviour in mice are mediated through its blockade of adenosine A(2A) receptors. *Br. J. Pharmacol.* **2000**, *129*, 1465–1473. [CrossRef]
59. Maximino, C.; Lima, M.G.; Olivera, K.R.M.; Picanço-Diniz, D.L.W.; Herculano, A.M. Adenosine A1, but not A2, receptor blockade increases anxiety and arousal in zebrafish. *Basic Clin. Pharmacol. Toxicol.* **2011**, *109*, 203–207. [CrossRef] [PubMed]
60. Yuan, M.; Chen, Y.; Huang, Y.; Lu, W. Behavioral and metabolic phenotype indicate personality in zebrafish (*Danio rerio*). *Front. Physiol.* **2018**, *9*, 653. [CrossRef] [PubMed]
61. Gutiérrez, H.C.; Vacca, I.; Schoenmacker, G.; Cleal, M.; Tochwin, A.; O'Connor, B.; Young, A.M.; Vasquez, A.A.; Winter, M.J.; Parker, M.O.; et al. Screening for drugs to reduce zebrafish aggression identifies caffeine and sildenafil. *Eur. Neuropsychopharmacol.* **2020**, *30*, 17–29. [CrossRef] [PubMed]
62. Yu, X.-D.; Zhu, Y.; Sun, Q.-X.; Deng, F.; Wan, J.; Zheng, D.; Gong, W.; Xie, S.-Z.; Shen, C.-J.; Fu, J.-Y.; et al. Distinct serotonergic pathways to the amygdala underlie separate behavioral features of anxiety. *Nat. Neurosci.* **2022**, *25*, 1651–1663. [CrossRef]
63. Baggio, S.; Mussulini, B.H.; de Oliveira, D.L.; Gerlai, R.; Rico, E.P. Embryonic alcohol exposure leading to social avoidance and altered anxiety responses in adult zebrafish. *Behav. Brain Res.* **2018**, *352*, 62–69. [CrossRef] [PubMed]
64. Capiotti, K.M.; Menezes, F.P.; Nazario, L.R.; Pohlmann, J.B.; de Oliveira, G.M.; Fazenda, L.; Bogo, M.R.; Bonan, C.D.; Da Silva, R.S. Early exposure to caffeine affects gene expression of adenosine receptors, DARPP-32 and BDNF without affecting sensibility and morphology of developing zebrafish (*Danio rerio*). *Neurotoxicol. Teratol.* **2011**, *33*, 680–685. [CrossRef] [PubMed]
65. Vida, C.; Montalvo, A.; García-Martín, M.; De la Fuente, M. A high neuroticism is accompanied by oxidative stress in peripheral blood cells of healthy young subjects. *Free Radic. Biol. Med.* **2018**, *120*, S103. [CrossRef]

toxics

MDPI

Article

The Role of Sex Steroid Hormones in the Association Between Manganese Exposure and Bone Mineral Density: National Health and Nutrition Examination Survey 2013–2018

Xiang Zhao [1,2,3,4,5,†], Jiayi Li [6,†], Jincong Yu [7], Yinhui Shi [8] and Mengling Tang [6,*]

[1] Department of Orthopaedics, Second Affiliated Hospital, Zhejiang University School of Medicine, Hangzhou 310009, China; flyingzhao@zju.edu.cn
[2] Orthopaedics Research Institute of Zhejiang University, Hangzhou 310009, China
[3] Key Laboratory of Motor System Disease Research and Precision Therapy of Zhejiang Province, Hangzhou 310009, China
[4] Clinical Research Center of Motor System Disease of Zhejiang Province, Hangzhou 310009, China
[5] State Key Laboratory of Transvascular Implantation Devices, Hangzhou 310009, China
[6] Department of Public Health, Fourth Affiliated Hospital, Zhejiang University School of Medicine, Hangzhou 310058, China
[7] Department of Orthopeadics, YuYao People's Hospital, Ningbo 315400, China
[8] Department of Orthopeadics, CHC International Hospital, Ningbo 315300, China
[*] Correspondence: tangml@zju.edu.cn; Tel.: +86-0571-88208194
[†] These authors contributed equally to this work.

Abstract: This study investigates the association between blood Mn and bone mineral density (BMD), focusing on the mediating role of sex steroids, using data from 8617 participants in the National Health and Nutrition Examination Survey (NHANES) 2013–2018. Weighted multiple linear regression models were used to examine the association of blood Mn and total BMD, and mediation analyses were used to explored the roles of total testosterone (TT), estradiol (E2), and sex hormone-binding globulin (SHBG) in the Mn-BMD relationship, stratified by sex and menopausal status. Blood Mn was negatively associated with BMD in both sexes, with a pronounced effect in postmenopausal women. SHBG mediated 37.16% of the Mn-BMD association in men, whereas no mediating effects were found in women. E2 exhibited a significant indirect effect, suggesting that reduced E2 levels may amplify Mn's effect on BMD. These findings indicate that Mn exposure is associated with decreased BMD, potentially through alterations in sex steroids, highlighting the importance of considering hormone status when evaluating the impact of Mn exposure on BMD.

Keywords: manganese; bone mineral density; sex steroid hormone; mediation; NHANES

1. Introduction

Osteoporosis is a disease characterized by decreased bone mineral density (BMD), compromised bone architecture, and an increased risk of fracture [1]. According to the World Health Organization (WHO), osteopenia is defined as a BMD value measured by dual-energy X-ray absorptiometry (DXA) that is 1 to 2.5 standard deviations below the mean of a healthy, sex-matched population. Osteoporosis is a significant risk factor for fractures, particularly in postmenopausal women and the elderly [2,3]. According to a study from the 2005–2010 National Health and Nutrition Examination Survey (NHANES), the prevalence of osteoporosis in U.S. adults aged 50 and older is 10.3%, with projections suggesting it could rise to 32% by 2030 [4]. Both intrinsic factors such as genetics, age,

https://doi.org/10.3390/toxics13040296

and sex, as well as extrinsic environmental considerations like exposure to metals may contribute to the risk of BMD loss [5,6].

Manganese (Mn) is a common heavy metal widely present in mines, steel mills, dry cell battery manufacturing plants, and welding operations [7]. Excessive exposure to Mn through inhalation, ingestion, or skin contact can lead to toxicity [8], which is associated with adverse health outcomes, such as impaired neurodevelopment [9], inflammatory liver damage [10], spleen apoptotic injury [11], and testes dysfunction [12]. Bone is one of the primary tissues that accumulates Mn, holding approximately 40% of the body's total Mn content [13]. Despite this, epidemiological studies on the relationship between Mn exposure and BMD are limited and yield mixed results. For example, several NHANES studies have reported negative correlations between blood Mn and BMD in both adolescents and adults [14–16]. Similarly, a study among retired Chinese workers found that 304 women with long-term Mn exposure had a higher incidence of osteoporosis compared to 277 controls [17]. In contrast, a study of 627 Chinese individuals aged 50 and older suggested that co-exposure to high concentrations of Mn, iron (Fe), copper (Cu), and selenium (Se) may have a protective effect on bone health, although Mn was not identified as the main contributor to the observed outcomes [7]. These contradictory findings may be due to differences in population size, age and sex composition, as well as in co-exposure to other metals, highlighting the need for further investigations into the effects of Mn exposure on bone health.

Previous studies have suggested that excessive Mn may interfere with bone metabolism by disrupting the endocrine system, altering sex steroid levels, and thereby increasing the risk of bone loss [18,19]. In vitro experiments have shown that Mn may inhibit sex steroid biosynthesis by reducing StAR protein levels, which are responsible for cholesterol transport to the mitochondria, where sex steroids are synthesized [20]. Sex steroid hormones play a key role in maintaining bone mass. Estrogen (E2) regulates bone remodeling by inhibiting bone resorption and promoting bone formation. Testosterone acts on bone either directly by stimulating androgen receptors, or indirectly by being converted into E2 by aromatase enzymes in peripheral tissues (e.g., adipose tissue, skin, bone, brain, and liver), or influencing bone metabolism through cytokines and growth factors [21,22]. Sex hormone-binding globulin (SHBG) regulates the activity of sex steroids by binding them. Nevertheless, relevant evidence is limited, and we believe that studying the role of sex steroids might help elucidate the link between Mn exposure and BMD loss.

As noted, we hypothesize that sex steroid hormones may mediate the effects of Mn exposure on BMD. Investigating the relationship between Mn, BMD, and sex steroids may help clarify the pathways through which Mn induces bone loss. Therefore, we conducted a cross-sectional study to examine the association between Mn exposure and BMD, as well as the mediating roles of sex steroid hormones using data from the NHANES.

2. Materials and Methods

2.1. Study Population

Data used in this study were collected from the 2013–2018 NHANES. The NHANES was a population-based program designed to assess the health and nutritional status of U.S. adults and children. The NHANES employed a stratified multistage probability cluster sampling and collected information through a series of in-home interviews and standardized physical examinations, which surveyed about 5000 persons per year. The National Center for Health Statistics of the Centers for Disease Control and Prevention's Research Ethics Review Board ratified the procedure and all participants provided informed consent.

Among three NHANES cycles (2013–2014, 2015–2016, and 2017–2018), 16,927 individuals aged 8–59 years and not pregnant underwent BMD measurement. From this group, those with missing data on either BMD (n = 3577) or blood Mn (n = 4733) were further excluded, leaving a total of 8617 participants. Due to the absence of serum sex steroid hormone measurements in the 2017–2018 cycle, we further focused on 4540 participants with complete information on total BMD, blood Mn, and sex steroid hormones from the two earlier cycles (2013–2014 and 2015–2016). After excluding individuals with a history of female hormone use (n = 101), those who had undergone oophorectomy (n = 22), and participants with missing covariates of BMI (n = 9) or cotinine (n = 3), a final total of 4405 participants were included for the mediation analysis. All participants were adults aged 20 years or over. The specific inclusion process of study subjects is presented in Figure 1.

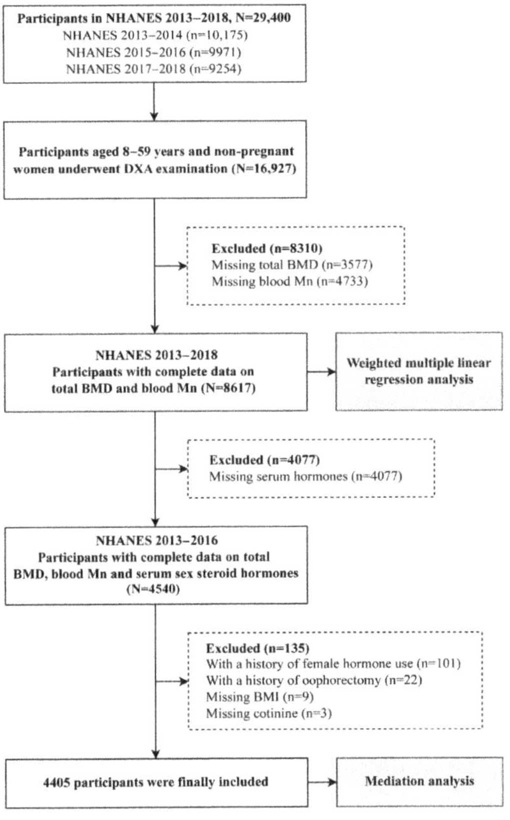

Figure 1. The inclusion process of the participants.

2.2. *Measurement of Variables*

The exposure variable was blood Mn, measured by an inductively coupled plasma mass spectrometer (ICP-MS) [23]. The lower limit of detection (LLOD) for blood Mn was 0.99 μg/L and the LLOD divided by the square root of two was used to indicate the measured values below LLOD. As the outcome variable, the total BMD was measured using DXA. The whole-body scans were performed using software version Apex 3.2 on the Hologic Discovery model A densitometers (Hologic, Inc., Bedford, Marlborough, MA, USA). Radiology technologists who performed the DXA examination were all well trained and certified. Less than 20 uSv of radiation was emitted during DXA whole-body scans, which was extremely minimal. Individuals aged < 8 or ≥60 years, and pregnant females (positive urine pregnancy test and/or self-report), were excluded from the DXA examination. BMD data were set to missing for participants who self-reported the use of radiographic contrast material (barium) within the past 7 days, or self-reported weight over 450 pounds or height over 6′5″.

Three typical sex steroid hormones, total testosterone (TT), E2, and SHBG, in serum were quantified in NHANES 2013–2014 and 2015–2016 cycles. Wherein TT and E2 were measured using isotope dilution liquid chromatography-tandem mass spectrometry (ID-LC/MS/MS) [24], and SHBG was quantified based on its reaction with immuno-antibodies and chemo-luminescence measurements of the reaction products occurring after two incubation periods. The LLODs for TT, E2, and SHBG were 0.75 ng/mL, 2.994 pg/mL, and 0.800 nmol/L, respectively.

2.3. Menopausal Status Definitions

Hormone levels varied among women based on their menopausal status. Therefore, we further separated the female sample in analyses into premenopausal and postmenopausal groups, as determined by a self-reported reproductive health questionnaire in NHANES. Women were classified as premenopausal if they responded "yes" to the question, "Have you had at least one menstrual period in the past 12 months?". Postmenopausal women were those who answered "hysterectomy" or "menopause/change of life" to the question "What is the reason that you have not had a period in the past 12 months?".

2.4. Covariates

Several potential confounders, identified from previous Mn-BMD studies and theoretical considerations, were incorporated into the analyses, including age (continuous in years), sex (male or female), race/ethnicity, body mass index (BMI), smoking status, and physical activity (PA). Race/ethnicity, as NHANES defined, was categorized as Mexican American, other Hispanic, non-Hispanic white, non-Hispanic black, and other races. BMI was calculated by dividing weight in kilograms by the square of height in meters (kg/m^2) and was categorized as follows: underweight/normal weight (<25.0 kg/m^2), overweight (25.0~29.9 kg/m^2), and obesity (≥30.0 kg/m^2). Smoking status, as a continuous variable, was determined by serum cotinine level (ng/mL) [25]. PA levels were assessed using the validated International Physical Activity Questionnaire (IPAQ) and calculated based on the metabolic equivalent (MET) values for the type, frequency, and duration of activities per week. The results were then categorized into three levels: <600 MET-min/week, 600~3000 MET-min/week, and >3000 MET-min/week.

2.5. Statistical Analysis

To account for the complex sampling design of NHANES, we used weighted proportions for categorical variables and weighted means (standard errors, SEs) for continuous variables to describe participant characteristics. Participants were categorized into four groups based on BMD quartiles. Differences in continuous variables across groups were assessed using the weighted linear regression, while categorical variables were compared using the Rao-Scott chi-square test. For the descriptive and regression analyses applied to all included participants, we created a 6-year weight as one-third of the value of the examination weights to represent the smallest subsample of the study.

Blood Mn and serum sex steroid hormones were log-transformed to approximate a normal distribution. To examine the potential non-linearity between blood Mn and total BMD, a restricted cubic spline (RCS) regression with 4 knots at the 5th, 35th, 65th, and 95th percentiles was applied. Based on the RCS results, weighted multiple linear regression models were used to investigate the association between blood Mn and total BMD [26], and effect estimates were presented as the percentage (%) change in total BMD per interquartile range (IQR) in blood Mn. Moreover, subgroup analyses were performed by age (<45 and ≥45 years), sex (male and female), BMI (<25.0, 25.0~29.9, and ≥30.0 kg/m^2), and PA levels (<600, 600~3000, and >3000 MET-min/week), and interaction tests were conducted to examine the Mn-BMD association across these subgroups. Three models were conducted as follows: Model 1 (crude model), Model 2 (adjusted for age, sex, and race/ethnicity), and Model 3 (adjusted for age, sex, race/ethnicity, BMI, smoking status and PA, with additional adjustment for menopause status in women). Mediation analyses was performed using the R mediation package [27]. One-half subsample blood metal subsample weights were used, and bootstrap analyses with 1000 resamples were performed. All data analyses were conducted using R software 4.2.2. A two-tailed p value < 0.05 was considered statistically significant.

2.6. Sensitivity Analysis

There is the possibility that the Mn–BMD association is attenuated by the impact of osteoporosis treatment or glucocorticoids use. Therefore, we excluded individuals who answered "yes" to the question "Have you ever been told by a doctor or other health care professional to take a prescribed medicine for osteoporosis?" (n = 60) and those who answered "yes" to the question "Have you ever taken any prednisone or cortisone pills nearly every day for a month or longer?" (n = 58) in NHANES 2013–2014, 2017–2018, but such information was not available in NHANES 2015–1016.

3. Results

3.1. Characteristics of the Study Population

Table 1 shows the characteristics of 8617 participants extracted from the NHANES 2013–2018. Nearly half of the participants were male (50.54% weighted), with a mean age of 32.49 ± 0.22 years. Significant differences were observed across the four groups based on BMD quartiles in terms of age, sex, race/ethnicity, BMI, smoking status, PA, blood Mn, serum TT, E2, and SHBG (all p < 0.001). Overall, participants with lower BMD exhibited significantly higher levels of blood Mn and serum SHBG, while having lower levels of serum TT and serum E2. Additionally, those with elevated BMD were more likely to be older, male, have a higher BMI, be smokers, and engage in higher levels of PA.

Table 1. Characteristics of the study participants by BMD quartiles in NHANES 2013–2018.

Variables	Quartile 1 (<0.931 g/cm^2)		Quartile 2 (0.931–1.056 g/cm^2)		Quartile 3 (1.057–0.1.144 g/cm^2)		Quartile 4 (≥1.145 g/cm^2)		*p* Value
	N	Mean (SE) or Percent (%)	*N*	Mean (SE) or Percent (%)	*N*	Mean (SE) or Percent (%)	*N*	Mean (SE) or Percent (%)	
Age (years)	2155	16.14 (0.77)	2154	33.59 (0.50)	2154	35.51 (0.38)	2154	37.15 (0.30)	<0.001
Sex, men	1069	49.04	785	35.00	1013	46.41	1412	67.91	<0.001
Race/ethnicity									<0.001
Mexican American	492	16.65	431	13.56	407	12.65	298	9.55	
Other Hispanic	247	8.82	268	9.72	227	7.84	182	6.56	
Non-Hispanic White	620	53.26	679	57.59	730	60.19	690	58.54	
Non-Hispanic Black	406	9.64	300	7.11	350	8.54	660	16.60	
Other race	390	11.62	476	12.02	440	10.78	324	8.75	
BMI (kg/m^2)									<0.001
<25.0	1770	79.15	1066	45.09	757	34.16	595	26.55	
25.0~29.9	255	13.78	565	27.91	627	29.49	653	33.13	
≥30.0	127	7.07	517	27.00	765	36.36	896	40.32	
Cotinine level (ng/mL)	2118	13.22 (2.27)	2126	46.09 (3.49)	2128	53.58 (4.82)	2135	64.66 (5.93)	<0.001
PA (MET-min/week)									<0.001
<600	1901	82.29	987	40.55	744	30.23	597	24.59	
600~3000	154	10.98	598	29.93	635	32.64	600	28.66	
>3000	100	6.73	569	29.51	775	37.13	957	46.75	
Mn (ng/mL)	2155	10.99 (0.14)	2154	10.71 (0.12)	2154	10.14 (0.11)	2154	9.60 (0.10)	<0.001
TT (ng/dl)	1285	57.53 (6.93)	1103	175.33 (8.1)	1089	220.23 (9.67)	1063	301.13 (7.88)	<0.001
E2 (pg/mL)	1285	17.15 (1.71)	1103	54.35 (2.77)	1089	52.32 (2.92)	1063	48.54 (2.71)	<0.001
SHBG (nmol/L)	1285	81.71 (3.16)	1103	60.23 (1.43)	1089	55.04 (1.90)	1063	48.16 (1.46)	<0.001

Notes: Sample size changes due to data availability. Abbreviations: BMD, bone mineral density; BMI, body mass index; PA, physical activity; Mn, manganese; TT, total testosterone; E2, estradiol; SHBG, sex hormone-binding globulin; MET, metabolic equivalent; SE, standard error.

3.2. Association Between Blood Mn and Total BMD

As shown in Figure 2, no significant non-linearity was detected between log-transformed blood Mn and total BMD using RCS. Table 2 reveals a significantly negative association between blood Mn and total BMD across all three models. After adjusting for all potential confounders, each IQR (4.46 ng/mL) increase in blood Mn was associated with a −9.3% (−13.5%, −4.9%) change in BMD.

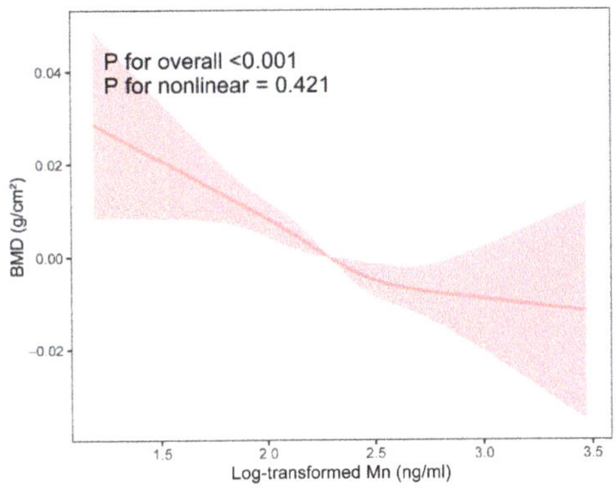

Figure 2. RCS analysis of the association between log-transformed blood Mn and total BMD.

Table 2. Association between blood Mn and total BMD (NHANES 2013–2018).

Model [a]		*n*	β (95% CI) [b]	% Change (95% CI) [c]	Mn, IQR (ng/mL)
Model 1	Total	8617	−0.065 (−0.077, −0.052) *	−24.9% (−28.9%, −20.7%) *	4.43
	Men	4279	−0.080 (−0.100, −0.060) *	−26.8% (−32.3%, −20.8%) *	3.91
	Women	4338	−0.024 (−0.0.37, −0.011) *	−11.1% (−16.6%, −5.2%) *	4.88
	Premenopausal	2445	−0.019 (−0.035, −0.004) *	−9.5% (−16.6%, −1.9%) *	5.22
	Postmenopausal	641	−0.043 (−0.066, −0.019) *	−17.0% (−25.0%, −8.1%) *	4.35
Model 2	Total	8617	−0.021 (−0.034, −0.009) *	−9.0% (−13.9%, −3.9%) *	4.43
	Men	4279	−0.044 (−0.061, −0.027) *	−15.8% (−21.3%, −9.9%) *	3.91
	Women	4338	−0.021 (−0.035, −0.008) *	−9.8% (−15.5%, −3.7%) *	4.88
	Premenopausal	2445	−0.015 (−0.030, 0.001)	−7.5% (−14.6%, 0.3%)	5.22
	Postmenopausal	641	−0.041 (−0.064, −0.020) *	−16.5% (−24.3%, −7.8%) *	4.35
Model 3	Total	7441	−0.022 (−0.033, −0.011) *	−9.3% (−13.5%, −4.9%) *	4.46
	Men	3675	−0.037 (−0.053, −0.021) *	−13.4% (−18.6%, −7.8%) *	3.88
	Women	3766	−0.027 (−0.040, −0.014) *	−12.4% (−17.7%, −6.7%) *	4.86
	Premenopausal	2168	−0.019 (−0.034, −0.004) *	−9.5% (−16.2%, −2.2%) *	5.23
	Postmenopausal	631	−0.045 (−0.068, −0.023) *	−17.9% (−25.6%, −9.6%) *	4.36

Notes: Sample size changes due to data availability. [a] Model 1 was a crude model; Model 2 adjusted for age and race/ethnicity; Model 3 adjusted for age, race/ethnicity, BMI, smoking status, and PA. In addition, sex was adjusted in Model 2 and 3 for the total population, and menopause status in all models for women. [b] Regression coefficient (95% CI) for a 1-unit increase in log-transformed blood Mn on BMD. [c] Percentage change (95% CI) in BMD for each IQR increase in blood Mn. * $p < 0.05$. Abbreviations: BMD, bone mineral density; Mn, manganese; BMI, body mass index; PA, physical activity; CI, confidence interval; IQR, interquartile range.

Similar negative associations were observed in both men and women. Further stratified analyses of the women group by menopause status revealed a stronger negative association between blood Mn and total BMD in postmenopausal women (% BMD change per 4.36 ng/mL Mn increase = −17.9%, 95% CI: −25.6%, −9.6%) compared to premenopausal women (% BMD change per 5.32 ng/mL Mn increase = −9.5%, 95% CI: −16.2%, −2.2%). After excluding participants who underwent osteoporosis treatment and glucocorticoids users, the Mn-BMD association remained stable in size, except in postmenopausal women, where the effect was slightly enhanced, showing a −20.5% (−28.3%, −12.0%) change in BMD per IQR (4.44 ng/mL) increase in blood Mn (Table S1). In addition, the association was more pronounced in young adults (<45 years) (p for interaction < 0.05). There were also indications of stronger associations between higher blood Mn levels and lower total BMD in individuals with lower BMI (<25.0 kg/m^2), and those with higher PA levels ($\geq$600 MET-min/week), although these associations did not reach statistical significance (Figure 3).

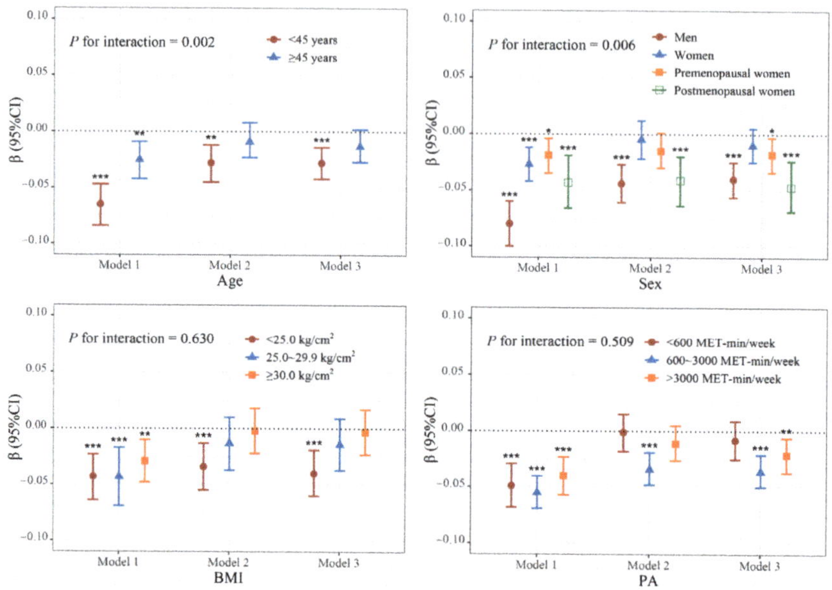

Figure 3. Subgroup analyses of the association between blood Mn and total BMD. *** $p < 0.001$, ** $p < 0.01$, * $p < 0.05$.

3.3. Association Between Blood Mn and Serum Sex Steroid Hormones

Table S2 demonstrates the associations of blood Mn with TT, E2, and SHBG in participants, and stratified by sex and menopausal status. In men, higher levels of blood Mn were associated with lower levels of TT, E2, and SHBG. However, after adjusting for all potential confounders, including age, race/ethnicity, BMI, smoking status, and PA, these associations were attenuated and became statistically non-significant, with the exception of SHBG, which remained negatively associated with blood Mn ($\beta = -0.130$, 95% CI: -0.220, -0.041). In women, no significant associations were observed between blood Mn and TT, E2, or SHBG, whether in the total women group or when stratified by menopausal status.

3.4. Association Between Serum Sex Steroid Hormones and Total BMD

Table S3 presents the associations of TT, E2, and SHBG with total BMD. In men, significant associations of TT, E2, and SHBG with BMD were observed. Specifically, TT and E2 were positively associated with BMD (TT: $\beta = 0.049$, 95% CI: 0.046, 0.053; E2: $\beta = 0.088$, 95% CI: 0.080, 0.096), whereas SHBG was negatively associated ($\beta = -0.055$, 95% CI: -0.065, -0.044). In women, only E2 showed a significant positive association with BMD, with a much smaller effect size than in men ($\beta = 0.008$, 95% CI: 0.004, 0.012), and this association was only observed in postmenopausal women.

3.5. Mediation Analyses

As shown in Table 3, the inverse relationship between blood Mn and total BMD was partially mediated by SHBG and E2. Specifically, the indirect effect of SHBG was 0.004 (95% CI: 0.002, 0.006) in the total population. This means that each IQR (4.47 ng/mL) increase in Mn was associated with lower SHBG, which in turn resulted in a 1.7% (1.0%, 2.6%) change in BMD, with a proportion of mediation of −15.70% (−33.05, −8.03). This effect was more substantial in men, where a per IQR (3.87 ng/mL) increase in Mn resulted in a 5.8% (4.2%, 7.6%) change in BMD, with a proportion of mediation of −37.16% (−66.07, −21.66). However, no significant mediating effects of TT, E2, or SHBG were observed in women. Although stratifying analyses by menopausal status revealed that E2 exhibited a significant indirect effect of -0.007 (95% CI: −0.021, −0.000) in postmenopausal women, the proportion of mediation did not reach statistical significance. Overall, the mediation effects of SHBG and E2 varied across different subgroups.

Table 3. Mediation of the association between blood Mn and total BMD by serum sex steroid hormones (n = 4405).

	Mn, IQR (ng/mL)	Mediator	Direct Effect			Indirect Effect			Proportion Mediated % (95% CI)
			β (95% CI) [a]	% Change (95% CI) [b]	p Value	β (95% CI)	% Change (95% CI)	p Value	
Total	4.47	TT	**−0.026 (−0.036, −0.016)**	**−10.9% (−14.8%, −6.3%)**	**<0.001**	0.001 (−0.002, 0.006)	0.7% (−1.4%, 2.6%)	0.492	−5.99 (−31.66, 9.23)
		E2	**−0.029 (−0.039, −0.020)**	**−12.1% (−15.8%, −8.4%)**	**<0.001**	0.005 (0.000, 0.009)	2.1% (−0.1%, 4.2%)	0.056	−18.91 (−54.92, −0.08)
		SHBG	**−0.028 (−0.039, −0.017)**	**−11.8% (−16.2%, −7.6%)**	**<0.001**	**0.004 (0.002, 0.006)**	**1.7% (1.0%, 2.6%)**	**<0.001**	**−15.70 (−33.05, −8.03)**
Men	3.87	TT	**−0.043 (−0.056, −0.029)**	**−15.2% (−19.4%, −10.6%)**	**<0.001**	0.004 (−0.004, 0.001)	1.4% (−1.4%, 4.0%)	0.298	−9.01 (−39.93, 7.56)
		E2	**−0.043 (−0.057, −0.029)**	**−15.3% (−19.5%, −10.8%)**	**<0.001**	0.004 (−0.003, 0.011)	1.4% (−1.2%, 4.3%)	0.306	−9.37 (−38.00, 6.98)
		SHBG	**−0.054 (−0.069, −0.039)**	**−18.7% (−23.2%, −14.2%)**	**<0.001**	**0.015 (0.010, 0.019)**	**5.8% (4.2%, 7.6%)**	**<0.001**	**−37.16 (−66.07, −21.66)**
Women	5.20	TT	**−0.029 (−0.043, −0.014)**	**−14.1% (−20.7%, −7.2%)**	**<0.001**	0.000 (−0.002, 0.001)	−0.2% (−0.8%, 0.3%)	0.452	1.07 (−1.75, 5.94)
		E2	**−0.029 (−0.043, −0.013)**	**−13.8% (−20.7%, −6.9%)**	**<0.001**	−0.001 (−0.003, 0.001)	−0.5% (−1.3%, 0.3%)	0.246	2.96 (−1.99, 10.14)
		SHBG	**−0.029 (−0.045, −0.013)**	**−14.1% (−20.4%, −7.3%)**	**<0.001**	0.000 (−0.001, 0.001)	−0.1% (−0.5%, 0.3%)	0.704	0.35 (−2.01, 3.86)
Premenopausal	5.37	TT	**−0.025 (−0.042, −0.010)**	**−12.5% (−19.7%, −5.2%)**	**<0.001**	−0.001 (−0.002, 0.000)	−0.4% (−1.3%, 0.1%)	0.158	3.27 (−1.09, 12.55)
		E2	**−0.026 (−0.041, −0.009)**	**−12.8% (−20.0%, −5.3%)**	**0.002**	0.000 (−0.001, 0.001)	−0.1% (−0.6%, 0.4%)	0.778	0.49 (−4.09, 5.39)
		SHBG	**−0.026 (−0.042, −0.010)**	**−12.8% (−19.9%, −4.8%)**	**0.002**	0.000 (−0.001, 0.000)	−0.1% (−0.5%, 0.3%)	0.750	0.44 (−2.12, 4.44)
Postmenopausal	4.16	TT	−0.034 (−0.076, 0.004)	−13.2% (−27.9%, 2.2%)	0.084	0.001 (−0.004, 0.006)	0.3% (−1.5%, 2.8%)	0.764	−2.01 (−45.69, 38.25)
		E2	−0.026 (−0.068, 0.016)	−10.3% (−24.6%, 6.7%)	0.222	**−0.007 (−0.021, −0.000)**	**−3.0% (−8.0%, −0.2%)**	**0.026**	21.90 (−71.78, 208.69)
		SHBG	−0.034 (−0.074, 0.007)	−13.1% (−25.8%, 2.6%)	0.102	0.000 (−0.003, 0.005)	0.2% (−1.3%, 1.8%)	0.082	−1.30 (−24.29, 19.85)

Notes: Models were adjusted for age, race/ethnicity, BMI, smoking status, and PA. In addition, sex was adjusted in the total population, and menopause status in women. [a] Regression coefficient (95% CI) for a 1-unit increase in log-transformed blood Mn on BMD. [b] Percentage change (95% CI) in BMD for each IQR increase in blood Mn. Bold indicates $p < 0.05$. Abbreviations: Mn, manganese; BMD, bone mineral density; TT, total testosterone; E2, estradiol; SHBG, sex hormone-binding globulin.

4. Discussion

4.1. Main Study Findings

Based on the 2013–2018 NHANES data, our study demonstrated a negative association between blood Mn and total BMD, which remained robust in both sexes. Notably, this negative correlation was more pronounced in postmenopausal women and among young adults. Furthermore, our studies revealed that SHBG and E2 mediated the association between blood Mn and total BMD across different subgroups. Specifically, SHBG mediated 37.16% of the association between blood Mn and total BMD among men, while no mediating effects were found in women. E2 in postmenopausal women exhibited a significant indirect effect on the relationship between blood Mn and total BMD. However, no significant mediating effect of TT was observed in either sex.

4.2. Interpretations and Comparisons with Other Studies

Previous studies have explored the effects of Mn exposure on bone mass, but the conclusions have been inconsistent. For instance, an animal study revealed that Mn supplementation might protect against ovariectomy-induced osteopenia in ovariectomized Sprague-Dawley rats fed a diet with varying Mn intake amounts [28]. In a Chinese population-based study of 51 seniors, cancellous bone Mn concentration was considerably higher in the non-osteoporosis group (1.96%) compared to the osteoporosis group (0.81%) [29]. However, a cross-sectional study analyzing data from 2545 adults in the NHANES 2011–2016 found a negative association between multiple metals co-exposure and BMD, with Mn being a primary contributor, accounting for 26.3% of the effect [16]. Another NHANES study involving 1703 U.S. adolescents also confirmed a negative association between blood Mn and BMD [14]. The discrepancies in findings across studies may stem from variations in study design, population characteristics, sample sizes, and the levels of Mn exposure. Our study, which supplements previous NHANES analyses with additional samples, similarly found a negative association between Mn and BMD in both sexes. Specifically, each IQR increase (around 4.0 ng/mL) in blood Mn was associated with a change in total BMD from −17.9% to −9.3% across different sample groups.

The exact mechanisms underlying this relationship remained unclear. Several potential explanations may apply. First, Mn functions as a cofactor for several enzymes. Excessive Mn may lead to enzymatic abnormalities that disrupt bone metabolism. For example, Mn superoxide dismutase could enhance the production and activity of osteoclasts [20], promoting thinning and degradation of trabecular bone, which eventually contributes to osteoporosis. Second, BMD reduction was intimately related to oxidative stress [18]. Mn exposure may induce osteotoxicity by increasing the production of reactive oxygen species (ROS), exacerbating oxidative stress and inflammation. Furthermore, excessive Mn could interfere with the metabolism of macro minerals and other trace elements, disrupting bone tissue [30].

Another potential mechanism involves sex steroids. In this study, Mn exposure was negatively correlated with serum SHBG levels, consistent with previous studies. A Chinese study in 118 men reported an inverse correlation between urinary Mn and TT levels [31]. Lower SHBG levels may reduce Mn's negative impact on BMD in men. Previous research reported the negative association between SHBG and osteoporosis [32]. SHBG regulates the bioavailability of sex hormones such as TT and E2 by binding to them, reducing their free, active form. Elevated SHBG levels may decrease the biological activity of TT in men, including its protective effect on bone. As known, TT performs a crucial role in maintaining bone mass, especially in men. TT in vivo can convert to dihydrotestosterone, inducing androgenic activity by binding to androgen receptors. TT also acts directly on osteoblasts by androgen receptors, promoting bone formation and increasing bone mass [22]. A clinical trial with 105 males with type 2 diabetes injected intramuscularly with testosterone cypionate (200 mg) biweekly for 18 months proved that testosterone therapy contributed to increased BMD [33]. Cross-sectional studies of postmenopausal women from the NHANES also reported positive associations between serum TT levels and BMD [34,35], indicating the protective effect of high TT levels on bone mass. Nevertheless, the mediation by TT in the relationship between Mn exposure and BMD was not evident in our findings.

The observed reduction in SHBG levels could represent a physiological response to Mn exposure, potentially serving as an adaptive mechanism to enhance the bioavailability of bone-protective sex hormones. Mn-induced oxidative stress may impair hepatocyte function, leading to reduced SHBG synthesis [36], or disrupt the hypothalamic–pituitary–gonadal (HPG) axis, thereby affecting SHBG regulation [37]. Additionally, Mn exposure may trigger systemic inflammation, increasing cytokine levels such as IL-6 and TNF-α, both of which have been reported to downregulate SHBG production [38]. Similar effects have been observed for other metals [39], For instance, cadmium (Cd) exposure was linked to alterations in the levels and activity of sex steroid hormone receptors, potentially influencing intracellular signaling and contributing to a proinflammatory state in endothelial cells [40]. These findings support the hypothesis that metal exposures, including Mn, may influence SHBG through common mechanisms. However, these hypotheses warrant further investigation.

Previous studies have explored the role of E2 in the relationship between heavy metal exposure and adverse health outcomes [41–43], though results have been inconsistent across different populations. In this study, the stronger negative association between Mn and BMD in postmenopausal women compared to premenopausal women suggests that lower E2 levels may exacerbate the effects of Mn exposure on bone health. Previous animal studies have shown that E2 can enhance the activity of antioxidant enzymes such as manganese superoxide dismutase (MnSOD), which helps mitigate oxidative damage exacerbated by Mn exposure, thus protecting bone cells. E2 may also interact with estrogen receptors (ERs) on mitochondria to help maintain mitochondrial homeostasis [44], preventing bone loss caused by mitochondrial dysfunction. Additionally, E2 supports bone mass by influencing osteoclast and osteoblast activity [45] and aiding in calcium absorption and retention [46]. As E2 levels decline after menopause, its protective role in bone metabolism weakens, increasing the vulnerability to bone loss from factors like Mn exposure. In contrast, premenopausal women, with more stable and higher E2 levels, may be better protected. However, the wide confidence intervals of mediation estimates suggest considerable uncertainty. These findings imply that while E2 may have a role in the relationship between Mn exposure and BMD, its exact contribution remains unclear and warrants further investigation.

We identified a more pronounced negative correlation between blood Mn and total BMD in young adults (aged < 45 years). During key stages of bone growth, particularly in younger individuals, the increase in bone density and the maturation of bone structure are crucial for lifelong bone health [47]. At this stage, bone turnover is faster, and bone remodeling is more active, making it more susceptible to external disruptions. Furthermore, younger individuals generally have a higher metabolic rate, which may lead to faster intake and accumulation of Mn, exacerbating its toxic effects on bone tissue. Therefore, exposure to Mn during this period of bone development could have a more significant negative impact on BMD.

Moreover, the negative association appeared to be more pronounced in individuals with lower BMI, suggesting that Mn's impact on BMD might be influenced by body composition. Lower BMI was often associated with reduced fat reserves, which may result in less protection for bones [48]. A pilot screening study among women in Singapore found that lower BMI was linked to an increased risk of low bone mass [49]. Adipose tissue plays a crucial role in bone metabolism, as fat cells secrete hormones like E2, which positively influence BMD [21,22]. Individuals with lower BMI tend to have relatively lower E2 levels [50], making their bones more susceptible to external factors such as Mn exposure. In addition, Mn plays a significant role in regulating lipid metabolism. Animal studies have shown that lean mice have considerably higher Mn concentrations in their bones compared to obese mice [51]. A cross-sectional study in China found that higher Mn intake was associated with a reduced risk of abdominal obesity in men [52]. Alternatively, adipose tissue may function as a reservoir for Mn, reducing its bioavailability and thereby lessening its impact on BMD. Lean individuals may have less adipose tissue to sequester Mn, potentially leading to higher bioavailability and increased oxidative stress. Nevertheless, this hypothesis requires further verification.

4.3. Strengths and Limitations

A strength of this study was the use of a large, nationally representative sample of the U.S. population, which confirmed the negative association between blood Mn and total BMD. The study further explored the mediating role of three hormones (TT, E2, and SHBG) between Mn exposure and BMD, testing the hypothesis that Mn exposure may influence hormone levels and consequently reduce BMD. In addition, the study separated the sample by sex, and within women, stratified by menopausal status. This allowed the identification of different hormonal mediators across subgroups, highlighting the importance of sex- and menopause-specific effects that may not have been detected in a pooled sample. By accounting for these differences, the study provided a more nuanced understanding of how sex steroids modulate the relationship between Mn exposure and BMD in different demographic groups.

Nevertheless, several limitations deserve consideration. First, TT rather than free testosterone, was measured in this study. Although free testosterone accounted for only 1~2% of circulating TT, it is the biologically active form [53]. Most testosterone was bound to SHBG, which may limit our ability to fully access the role of testosterone in mediating the relationship between Mn exposure and BMD. Second, the relatively small sample size of women with menopausal information posed a challenge, particularly when stratifying by menopausal status, which limited the power of statistical testing. Third, strict homeostatic regulation, a short half-life, and individual variability limited the use of blood Mn as an indicator of long-term exposure [54]. However, we used blood Mn as an exposure indicator since the NHANES merely measured Mn content in whole blood and urine. Additionally, while we lacked data on specific exposure routes, which could influence Mn's bioavailability and toxicity, blood Mn levels reflect the integrated exposure from various pathways, including inhalation and ingestion. Fourth, the analysis was constrained by its cross-sectional design and therefore could not infer a causal relationship between blood Mn and BMD. Also, since this study primarily focused on the effects of Mn exposure on BMD, the potential synergistic or antagonistic effects of multiple metal exposures were beyond its scope. Indeed, other metals may also be risk factors for osteopenia or osteoporosis, such as lead (Pb), aluminum (Al), and Cd [55]. Given that metal exposures often occur simultaneously, future studies should investigate how co-exposure to various metals may affect bone health, and how these effects compare with our findings. Due to the lack of thyroid function data in the NHANES 2013–2018, we were unable to further explore the potential impact of conditions such as hypothyroidism and hyperparathyroidism on the Mn-BMD association. The biological mechanisms underlying our findings remained unclear, including whether interventions such as chelation therapy or hormone supplementation could mitigate Mn-related bone loss. Thus, further prospective studies, as well as experiments in vivo and in vitro, are needed to verify our findings, elucidate the physiological processes involved, and explore possible preventive strategies.

5. Conclusions

In this study, we found that exposure to Mn was negatively associated with total BMD, with sex steroids including SHBG and E2 serving as mediators. Specifically, SHBG mitigated the negative relationship between Mn exposure and BMD in men. Mn exposure was associated with decreased E2 levels, which in turn correlated with reduced BMD in postmenopausal women. These findings suggested that Mn exposure may disrupt hormonal balance, leading to decreased BMD, particularly in sex- and menopause-specific subgroups. Future research with larger sample size, as well as prospective cohort studies and experimental research, are needed to confirm these findings.

Supplementary Materials: The following supporting information can be downloaded at: https://www.mdpi.com/article/10.3390/toxics13040296/s1. Table S1 Sensitivity analyses of the association between blood Mn and total BMD: excluding individuals under osteoporosis treatment, and glucocorticoid users (NHANES 2013–2018). Table S2 Weighted regression coefficients for log-transformed serum sex steroid hormones relative to a unit increase in log-transformed blood Mn (NHANES 2013–2016). Table S3 Weighted regression coefficients for total BMD relative to a unit increase in log-transformed serum sex steroid hormones (NHANES 2013–2016).

Author Contributions: X.Z.: Conceptualization, methodology, software, formal analysis, writing—original draft preparation; J.L.: methodology, software, visualization, validation, writing—review and editing; J.Y.: methodology, data curation; Y.S.: methodology, data curation; M.T.: writing—review and editing, supervision, project administration. X.Z. and J.L. contributed equally to this manuscript. All authors have read and agreed to the published version of the manuscript.

Funding: This research received no external funding.

Institutional Review Board Statement: This study was conducted in accordance with the Declaration of Helsinki, and approved by the US National Center for Health Statistics (NCHS) Ethics Review Board (ERB) authorized the 2013—2018 NHANES (protocol number: Protocol #2011-17, Protocol #2018-01) (see details at https://www.cdc.gov/nchs/nhanes/irba98.htm (accessed on 24 February 2025)).

Informed Consent Statement: Not applicable.

Data Availability Statement: The data used in this study can be downloaded for free in NHANES (https://wwwn.cdc.gov/nchs/nhanes/Default.aspx (accessed on 24 February 2025)).

Acknowledgments: The authors would like to express their sincere gratitude to all the participants and staff of the NHANES study.

Conflicts of Interest: The authors declare no conflicts of interest.

References

1. Consensus, A. Consensus Development Conference: Diagnosis, Prophylaxis, and Treatment of Osteoporosis. *Am. J. Med.* **1993**, *94*, 646–650. [CrossRef]
2. Xiao, P.-L.; Cui, A.-Y.; Hsu, C.-J.; Peng, R.; Jiang, N.; Xu, X.-H.; Ma, Y.-G.; Liu, D.; Lu, H.-D. Global, Regional Prevalence, and Risk Factors of Osteoporosis According to the World Health Organization Diagnostic Criteria: A Systematic Review and Meta-Analysis. *Osteoporos. Int.* **2022**, *33*, 2137–2153. [CrossRef] [PubMed]
3. Yoo, J.E.; Shin, D.W.; Han, K.; Kim, D.; Yoon, J.W.; Lee, D.-Y. Association of Female Reproductive Factors with Incidence of Fracture Among Postmenopausal Women in Korea. *JAMA Netw. Open* **2021**, *4*, e2030405. [CrossRef]
4. Wright, N.C.; Looker, A.C.; Saag, K.G.; Curtis, J.R.; Delzell, E.S.; Randall, S.; Dawson-Hughes, B. The Recent Prevalence of Osteoporosis and Low Bone Mass in the United States Based on Bone Mineral Density at the Femoral Neck or Lumbar Spine. *J. Bone Min. Res.* **2014**, *29*, 2520–2526. [CrossRef]
5. Morris, J.A.; Kemp, J.P.; Youlten, S.E.; Laurent, L.; Logan, J.G.; Chai, R.C.; Vulpescu, N.A.; Forgetta, V.; Kleinman, A.; Mohanty, S.T.; et al. An Atlas of Genetic Influences on Osteoporosis in Humans and Mice. *Nat. Genet.* **2019**, *51*, 258–266. [CrossRef]

6. Huang, Z.; Wang, X.; Wang, H.; Zhang, S.; Du, X.; Wei, H. Relationship of Blood Heavy Metals and Osteoporosis among the Middle-Aged and Elderly Adults: A Secondary Analysis from NHANES 2013 to 2014 and 2017 to 2018. *Front. Public Health* **2023**, *11*, 1045020. [CrossRef] [PubMed]
7. Wei, M.; Huang, Q.; Dai, Y.; Zhou, H.; Cui, Y.; Song, W.; Di, D.; Zhang, R.; Li, C.; Wang, Q.; et al. Manganese, Iron, Copper, and Selenium Co-Exposure and Osteoporosis Risk in Chinese Adults. *J. Trace Elem. Med. Biol.* **2022**, *72*, 126989. [CrossRef]
8. Baj, J.; Flieger, W.; Barbachowska, A.; Kowalska, B.; Flieger, M.; Forma, A.; Teresiński, G.; Portincasa, P.; Buszewicz, G.; Radzikowska-Büchner, E.; et al. Consequences of Disturbing Manganese Homeostasis. *Int. J. Mol. Sci.* **2023**, *24*, 14959. [CrossRef]
9. Chung, S.E.; Cheong, H.-K.; Ha, E.-H.; Kim, B.-N.; Ha, M.; Kim, Y.; Hong, Y.-C.; Park, H.; Oh, S.-Y. Maternal Blood Manganese and Early Neurodevelopment: The Mothers and Children's Environmental Health (MOCEH) Study. *Environ. Health Perspect.* **2015**, *123*, 717–722. [CrossRef]
10. Liu, Y.; Yu, M.; Cui, J.; Du, Y.; Teng, X.; Zhang, Z. Heat Shock Proteins Took Part in Oxidative Stress-Mediated Inflammatory Injury via NF-κB Pathway in Excess Manganese-Treated Chicken Livers. *Ecotoxicol. Environ. Saf.* **2021**, *226*, 112833. [CrossRef]
11. Zhu, Y.; Li, S.; Teng, X. The Involvement of the Mitochondrial Pathway in Manganese-Induced Apoptosis of Chicken Splenic Lymphocytes. *Chemosphere* **2016**, *153*, 462–470. [CrossRef] [PubMed]
12. Du, Y.; Zhu, Y.; Teng, X.; Zhang, K.; Teng, X.; Li, S. Toxicological Effect of Manganese on NF-κB/iNOS-COX-2 Signaling Pathway in Chicken Testes. *Biol. Trace Elem. Res.* **2015**, *168*, 227–234. [CrossRef] [PubMed]
13. O'Neal, S.L.; Hong, L.; Fu, S.; Jiang, W.; Jones, A.; Nie, L.H.; Zheng, W. Manganese Accumulation in Bone Following Chronic Exposure in Rats: Steady-State Concentration and Half-Life in Bone. *Toxicol. Lett.* **2014**, *229*, 93–100. [CrossRef]
14. Liu, J.; Tang, Y.; Chen, Y.; Zhang, X.; Xia, Y.; Geng, B. Association between Blood Manganese and Bone Mineral Density in US Adolescents. *Env. Sci. Pollut. Res.* **2023**, *30*, 29743–29754. [CrossRef]
15. Wang, C.; Zhu, Y.; Long, H.; Ou, M.; Zhao, S. Relationship between Blood Manganese and Bone Mineral Density and Bone Mineral Content in Adults: A Population-Based Cross-Sectional Study. *PLoS ONE* **2022**, *17*, e0276551. [CrossRef]
16. Wei, M.-H.; Cui, Y.; Zhou, H.-L.; Song, W.-J.; Di, D.-S.; Zhang, R.-Y.; Huang, Q.; Liu, J.-A.; Wang, Q. Associations of Multiple Metals with Bone Mineral Density: A Population-Based Study in US Adults. *Chemosphere* **2021**, *282*, 131150. [CrossRef]
17. Li, D.; Ge, X.; Liu, Z.; Huang, L.; Zhou, Y.; Liu, P.; Qin, L.; Lin, S.; Liu, C.; Hou, Q.; et al. Association between Long-Term Occupational Manganese Exposure and Bone Quality among Retired Workers. *Environ. Sci. Pollut. Res. Int.* **2020**, *27*, 482–489. [CrossRef]
18. Li, L.; Yang, X. The Essential Element Manganese, Oxidative Stress, and Metabolic Diseases: Links and Interactions. *Oxidative Med. Cell. Longev.* **2018**, *2018*, 7580707. [CrossRef] [PubMed]
19. Zofková, I.; Nemcikova, P.; Matucha, P. Trace Elements and Bone Health. *Clin. Chem. Lab. Med. (CCLM)* **2013**, *51*, 1555–1561. [CrossRef]
20. Guo, T.; Zhang, L.; Konermann, A.; Zhou, H.; Jin, F.; Liu, W. Manganese Superoxide Dismutase Is Required to Maintain Osteoclast Differentiation and Function under Static Force. *Sci. Rep.* **2015**, *5*, 8016. [CrossRef]
21. David, K.; Narinx, N.; Antonio, L.; Evenepoel, P.; Claessens, F.; Decallonne, B.; Vanderschueren, D. Bone Health in Ageing Men. *Rev. Endocr. Metab. Disord.* **2022**, *23*, 1173–1208. [CrossRef] [PubMed]
22. Shigehara, K.; Izumi, K.; Kadono, Y.; Mizokami, A. Testosterone and Bone Health in Men: A Narrative Review. *J. Clin. Med.* **2021**, *10*, 530. [CrossRef]
23. Spaur, M.; Nigra, A.E.; Sanchez, T.R.; Navas-Acien, A.; Lazo, M.; Wu, H.-C. Association of Blood Manganese, Selenium with Steatosis, Fibrosis in the National Health and Nutrition Examination Survey, 2017–2018. *Environ. Res.* **2022**, *213*, 113647. [CrossRef]
24. Zhou, H.; Wang, Y.; Gatcombe, M.; Farris, J.; Botelho, J.C.; Caudill, S.P.; Vesper, H.W. Simultaneous Measurement of Total Estradiol and Testosterone in Human Serum by Isotope Dilution Liquid Chromatography Tandem Mass Spectrometry. *Anal. Bioanal. Chem.* **2017**, *409*, 5943–5954. [CrossRef]
25. Seccareccia, F.; Zuccaro, P.; Pacifici, R.; Meli, P.; Pannozzo, F.; Freeman, K.M.; Santaquilani, A.; Giampaoli, S. Serum Cotinine as a Marker of Environmental Tobacco Smoke Exposure in Epidemiological Studies: The Experience of the MATISS Project. *Eur. J. Epidemiol.* **2003**, *18*, 487–492. [CrossRef]
26. Zhang, M.; Hou, Y.; Ren, X.; Cai, Y.; Wang, J.; Chen, O. Association of a Body Shape Index with Femur Bone Mineral Density among Older Adults: NHANES 2007–2018. *Arch. Osteoporos.* **2024**, *19*, 63. [CrossRef] [PubMed]
27. Tofighi, D.; MacKinnon, D.P. RMediation: An R Package for Mediation Analysis Confidence Intervals. *Behav. Res.* **2011**, *43*, 692–700. [CrossRef] [PubMed]
28. Bae, Y.-J.; Kim, M.-H. Manganese Supplementation Improves Mineral Density of the Spine and Femur and Serum Osteocalcin in Rats. *Biol. Trace Elem. Res.* **2008**, *124*, 28–34. [CrossRef]
29. Lin, S.; Yang, F.; Ling, M.; Fan, Y. Association between Bone Trace Elements and Osteoporosis in Older Adults: A Cross-Sectional Study. *Ther. Adv. Musculoskelet. Dis.* **2022**, *14*, 1759720X221125984. [CrossRef]
30. Gaffney-Stomberg, E. The Impact of Trace Minerals on Bone Metabolism. *Biol. Trace Elem. Res.* **2019**, *188*, 26–34. [CrossRef]

31. Zeng, Q.; Zhou, B.; Feng, W.; Wang, Y.-X.; Liu, A.-L.; Yue, J.; Li, Y.-F.; Lu, W.-Q. Associations of Urinary Metal Concentrations and Circulating Testosterone in Chinese Men. *Reprod. Toxicol.* **2013**, *41*, 109–114. [CrossRef] [PubMed]
32. Narinx, N.; David, K.; Walravens, J.; Vermeersch, P.; Claessens, F.; Fiers, T.; Lapauw, B.; Antonio, L.; Vanderschueren, D. Role of Sex Hormone-Binding Globulin in the Free Hormone Hypothesis and the Relevance of Free Testosterone in Androgen Physiology. *Cell. Mol. Life Sci.* **2022**, *79*, 543. [CrossRef] [PubMed]
33. Colleluori, G.; Aguirre, L.; Napoli, N.; Qualls, C.; Villareal, D.T.; Armamento-Villareal, R. Testosterone Therapy Effects on Bone Mass and Turnover in Hypogonadal Men with Type 2 Diabetes. *J. Clin. Endocrinol. Metab.* **2021**, *106*, e3058–e3068. [CrossRef] [PubMed]
34. Yang, J.; Kong, G.; Yao, X.; Zhu, Z. Association between Serum Total Testosterone Level and Bone Mineral Density in Middle-Aged Postmenopausal Women. *Int. J. Endocrinol.* **2022**, *2022*, 4228740. [CrossRef]
35. Nunes, E.; Gallardo, E.; Morgado-Nunes, S.; Fonseca-Moutinho, J. Steroid Hormone Levels and Bone Mineral Density in Women over 65 Years of Age. *Sci. Rep.* **2023**, *13*, 4925. [CrossRef]
36. Sun, Y.; Li, S.; Liu, H.; Bai, H.; Hu, K.; Zhang, R.; Liu, Q.; Fan, P. Oxidative Stress Promotes Hyperandrogenism by Reducing Sex Hormone-Binding Globulin in Polycystic Ovary Syndrome. *Fertil. Steril.* **2021**, *116*, 1641–1650. [CrossRef]
37. Nkpaa, K.W.; Amadi, B.A.; Adedara, I.A.; Wegwu, M.O.; Farombi, E.O. Ethanol Exacerbates Manganese—Induced Functional Alterations along the Hypothalamic-Pituitary-Gonadal Axis of Male Rats. *Neurosci. Lett.* **2018**, *684*, 47–54. [CrossRef]
38. Osmancevic, A.; Daka, B.; Michos, E.D.; Trimpou, P.; Allison, M. The Association between Inflammation, Testosterone and SHBG in Men: A Cross-Sectional Multi-Ethnic Study of Atherosclerosis. *Clin. Endocrinol.* **2023**, *99*, 190–197. [CrossRef]
39. Chen, C.; Wang, N.; Nie, X.; Han, B.; Li, Q.; Chen, Y.; Zhai, H.; Zhu, C.; Chen, Y.; Xia, F.; et al. Blood Cadmium Level Associates with Lower Testosterone and Sex Hormone-Binding Globulin in Chinese Men: From SPECT-China Study, 2014. *Biol. Trace Elem. Res.* **2016**, *171*, 71–78. [CrossRef]
40. Fittipaldi, S.; Bimonte, V.M.; Soricelli, A.; Aversa, A.; Lenzi, A.; Greco, E.A.; Migliaccio, S. Cadmium Exposure Alters Steroid Receptors and Proinflammatory Cytokine Levels in Endothelial Cells in Vitro: A Potential Mechanism of Endocrine Disruptor Atherogenic Effect. *J. Endocrinol. Investig.* **2019**, *42*, 727–739. [CrossRef]
41. Chen, Y.; Pu, Y.; Liu, H.; Cao, A.; Du, Y.; He, S.; Ai, S.; Dang, Y. A Study on the Mediating Role of Serum Hormones in the Effects of Heavy Metals on Preeclampsia. *Environ. Pollut.* **2024**, *360*, 124721. [CrossRef]
42. Liu, H.; Li, Z.; Xie, L.; Jing, G.; Liang, W.; He, J.; Dang, Y. The Relationship Between Heavy Metals and Missed Abortion: Using Mediation of Serum Hormones. *Biol. Trace Elem. Res.* **2024**, *202*, 3401–3412. [CrossRef]
43. Li, X.; Yu, X.; Luo, K.; Liu, H.; Fan, X.; Yin, X.; Zhao, Q.; Liu, X.; Yang, Y. Exposure to Metals and the Disruption of Sex Hormones in 6–19 Years Old Children: An Exploration of Mixture Effects. *Ecotoxicol. Environ. Saf.* **2023**, *250*, 114477. [CrossRef] [PubMed]
44. Vasconsuelo, A.; Pronsato, L.; Ronda, A.C.; Boland, R.; Milanesi, L. Role of 17β-Estradiol and Testosterone in Apoptosis. *Steroids* **2011**, *76*, 1223–1231. [CrossRef] [PubMed]
45. Riggs, B.L.; Khosla, S.; Melton, L.J., III. Sex Steroids and the Construction and Conservation of the Adult Skeleton. *Endocr. Rev.* **2002**, *23*, 279–302. [CrossRef]
46. Pinsino, A.; Roccheri, M.C.; Costa, C.; Matranga, V. Manganese Interferes with Calcium, Perturbs ERK Signaling, and Produces Embryos with No Skeleton. *Toxicol. Sci.* **2011**, *123*, 217–230. [CrossRef]
47. Yang, Y.; Wu, F.; Winzenberg, T.; Jones, G. Tracking of Areal Bone Mineral Density from Age Eight to Young Adulthood and Factors Associated With Deviation From Tracking: A 17-Year Prospective Cohort Study. *J. Bone Miner. Res.* **2018**, *33*, 832–839. [CrossRef] [PubMed]
48. Lin, S.; Chen, C.; Cai, X.; Yang, F.; Fan, Y. The Concentrations of Bone Calcium, Phosphorus and Trace Metal Elements in Elderly Patients with Intertrochanteric Hip Fractures. *Front. Endocrinol.* **2022**, *13*, 1005637. [CrossRef]
49. Ang, S.B.; Xia, J.Y.; Cheng, S.J.; Chua, M.T.; Goh, L.; Dhaliwal, S.S. A Pilot Screening Study for Low Bone Mass in Singaporean Women Using Years since Menopause and BMI. *Climacteric* **2022**, *25*, 163–169. [CrossRef]
50. Qiao, D.; Li, Y.; Liu, X.; Zhang, X.; Qian, X.; Zhang, H.; Zhang, G.; Wang, C. Association of Obesity with Bone Mineral Density and Osteoporosis in Adults: A Systematic Review and Meta-Analysis. *Public Health* **2020**, *180*, 22–28. [CrossRef]
51. Kennedy, M.L.; Failla, M.L.; Smith, J.C. Influence of Genetic Obesity on Tissue Concentrations of Zinc, Copper, Manganese and Iron in Mice. *J. Nutr.* **1986**, *116*, 1432–1441. [CrossRef] [PubMed]
52. Zhou, B.; Su, X.; Su, D.; Zeng, F.; Wang, M.H.; Huang, L.; Huang, E.; Zhu, Y.; Zhao, D.; He, D.; et al. Dietary Intake of Manganese and the Risk of the Metabolic Syndrome in a Chinese Population. *Br. J. Nutr.* **2016**, *116*, 853–863. [CrossRef] [PubMed]
53. Kanakis, G.A.; Tsametis, C.P.; Goulis, D.G. Measuring Testosterone in Women and Men. *Maturitas* **2019**, *125*, 41–44. [CrossRef]
54. Karyakina, N.A.; Shilnikova, N.; Farhat, N.; Ramoju, S.; Cline, B.; Momoli, F.; Mattison, D.; Jensen, N.; Terrell, R.; Krewski, D. Biomarkers for Occupational Manganese Exposure. *Crit. Rev. Toxicol.* **2022**, *52*, 636–663. [CrossRef] [PubMed]
55. Battistini, B.; Greggi, C.; Visconti, V.V.; Albanese, M.; Messina, A.; De Filippis, P.; Gasperini, B.; Falvino, A.; Piscitelli, P.; Palombi, L.; et al. Metals Accumulation Affects Bone and Muscle in Osteoporotic Patients: A Pilot Study. *Environ. Res.* **2024**, *250*, 118514. [CrossRef]

toxics

MDPI

Article

Health Risk Assessment of Potentially Toxic Element Uptake by Lotus (*Nelumbo nucifera*) in Floating Lake Gardens

Mohssen Elbagory [1], Farahat S. Moghanm [2], Ibrahim Mohamed [3], Sahar El-Nahrawy [4], Alaa El-Dein Omara [4], Madhumita Goala [5], Pankaj Kumar [6,7], Boro Mioč [8], Željko Andabaka [8] and Ivan Širić [8,*]

1 Health Specialties, Basic Sciences and Applications Unit, Applied College, King Khalid University, Mohayil Asir Abha 61421, Saudi Arabia; mhmohammad@kku.edu.sa
2 Soil and Water Department, Faculty of Agriculture, Kafrelsheikh University, Kafr El-Sheikh 33516, Egypt; fsaadr@yahoo.ca
3 Department of Soil and Water Sciences, Faculty of Agriculture, Benha University, Moshtohor, Toukh 13736, Egypt; ibrahim.ali@fagr.bu.edu.eg
4 Soil Microbiology Research Department, Soils, Water, and Environment Research Institute (SWERI), Agriculture Research Center (ARC), Giza 12112, Egypt; sahar.elnahrawy@yahoo.com (S.E.-N.); alaa.omara@yahoo.com (A.E.-D.O.)
5 Department of Environmental Science, Graphic Era (Deemed to be University), Dehradun 248002, India
6 School of Environmental Studies, Maa Shakumbhari University, Punwarka, Saharanpur 247120, India
7 Research and Development Division, Society for AgroEnvironmental Sustainability, Dehradun 248007, India
8 Faculty of Agriculture, University of Zagreb, Svetosimunska 25, 10000 Zagreb, Croatia; bmioc@agr.hr (B.M.)
* Correspondence: isiric@agr.hr

Abstract: The present study investigated the uptake and health risks of potentially toxic elements (PTEs) by lotus (*Nelumbo nucifera*) cultivated in floating lake gardens of Dal Lake, Srinagar, India. Rapid urbanization and anthropogenic activities have led to PTE contamination in the lake, raising concerns about food safety and ecological sustainability. The objectives were to evaluate the physicochemical properties of water and sediment and to quantify PTEs (Cd, Cu, Cr, Co, Fe, Mn, Ni, and Zn) accumulation in different tissues of *N. nucifera* with associated health risks. A systematic sampling approach was adopted across four zones of the lake, collecting water, sediment, and plant tissues (August to October 2024). The results showed significant PTE contamination, with sediment showing high concentrations of Fe (1610.51 mg/kg), Mn (31.48 mg/kg), and Cr (29.72 mg/kg). Bioaccumulation factors indicated preferential PTE accumulation in roots, with Fe exhibiting the highest uptake (95.11). Translocation factors were low for most PTEs, suggesting limited mobility to edible parts. Health risk assessment indicated that Cr posed the highest non-carcinogenic risk (HRI = 1.8000 in rhizomes). The cumulative target hazard quotient (THQ) remained below 1, suggesting no immediate risk, but long-term exposure warrants concern. The study provided valuable information on the necessity of continuous monitoring and pollution mitigation strategies to ensure the food safety of floating lake garden-derived crops.

Keywords: bioaccumulation; floating lake gardens; health risk assessment; lotus cultivation; toxic element contamination

1. Introduction

The lotus (*Nelumbo nucifera*), commonly known as the sacred lotus, is an aquatic plant with deep evolutionary and ecological significance [1]. Fossil records suggest that *N. nucifera* has existed for over 135 million years, with its lineage dating back to the Cretaceous period [2]. Its distribution is largely affected by precipitation and availability of water bodies [3,4]. The species has developed specialized adaptations, such as hydrophobic leaves, thermoregulation in flowers, and seed dormancy, which allow it to thrive in aquatic environments [5]. *N. nucifera* plays an important role in stabilizing wetland ecosystems by improving water quality, controlling algal blooms, and providing habitat for various aquatic organisms [6,7]. Apart from its ecological importance, *N. nucifera* has been widely used in traditional medicine, culinary practices, and religious ceremonies [8]. The rhizomes, seeds, leaves, and flowers of *N. nucifera* contain bioactive compounds such as alkaloids, flavonoids, and polyphenols, which exhibit antioxidant, anti-inflammatory, and antimicrobial properties [9]. *N. nucifera* is valued in Asian cultures for its nutritional [10] and medicinal properties, used in Ayurveda and traditional Chinese medicine for various ailments, and symbolizes purity and enlightenment in Hinduism and Buddhism [11]. Thus, due to its wide ecological, medicinal, and cultural applications, *N. nucifera* is one of the significant aquatic plants in both natural and human-influenced landscapes [12].

N. nucifera horticulture practice is widespread across Asia, particularly in China, India, Japan, and Southeast Asian countries, where it is grown for food, medicine, and ornamental purposes [13,14]. China is the leading producer, cultivating *N. nucifera* on large-scale aquatic farms, particularly in provinces like Hubei and Jiangxi, where the plant is used for seed and rhizome production [15]. In India, *N. nucifera* is grown in the water bodies of Uttar Pradesh, Bihar, Kashmir, and Tamil Nadu states [16]. The edible parts of *N. nucifera*, especially the rhizomes (locally called *Nadru* in Kashmir) and seeds (*Makhana*), contribute significantly to local economies [17]. The growing market for *N. nucifera* products is related to herbal teas, essential oils, and pharmaceuticals [18], alongside its cultivation for religious, aesthetic, and agro-tourism purposes [19]. Therefore, the increasing recognition of *N. nucifera* as a high-value crop has encouraged research into optimized cultivation methods, value-added processing, and market expansion to enhance its commercial viability.

Floating lake gardens are a unique agricultural system practiced in various parts of the world that provides a sustainable means of cultivating aquatic crops like *N. nucifera* [20]. These gardens are prevalent in regions with extensive water bodies such as Dal Lake in Kashmir, Inle Lake in Myanmar, and parts of Bangladesh [21,22]. Farmers construct floating beds using organic materials such as decomposed plant matter, allowing crops to grow on water surfaces. This method optimizes land use, provides livelihood opportunities, and ensures year-round cultivation. In Dal Lake, floating gardens (*Rad*) serve as an essential resource for local farmers who grow *N. nucifera* alongside vegetables and other aquatic crops [23]. The system also enhances nutrient recycling in lakes, contributing to ecological balance. However, increasing anthropogenic activities have led to severe water quality degradation in many lakes, including Dal Lake [24]. Unregulated sewage discharge, agricultural runoff, and urban expansion have introduced high levels of organic pollutants, potentially toxic elements (PTEs), and excess nutrients, leading to eutrophication and contamination of aquatic ecosystems [25,26]. Water quality assessments of Dal Lake have revealed elevated concentrations of biochemical oxygen demand (BOD), chemical oxygen demand (COD), and potentially toxic elements (PTEs) such as cadmium (Cd), lead (Pb), and chromium (Cr) [27,28]. These pollutants pose serious threats to aquatic biodiversity, disrupt ecosystem functions, and jeopardize the safety of crops grown in floating lake gardens [29].

The ability of aquatic plants to absorb and accumulate PTEs raises concerns about the safety of edible crops cultivated in polluted water bodies [30]. *N. nucifera*, with its extensive root system, can absorb and store PTEs from sediments and water, leading to bioaccumulation in rhizomes, leaves, and seeds [31]. Studies have reported high concentrations of PTEs such as Cd, Pb, nickel (Ni), and arsenic (As) in *N. nucifera* tissues when grown in contaminated water [32]. The translocation of these PTE from roots to edible parts increases potential health risks for consumers. PTE accumulation in food crops is associated with several toxicological effects, including neurotoxicity, kidney damage, and carcinogenicity [33]. Chronic exposure to Cd and Pb through dietary intake can result in bioaccumulation in human tissues, leading to oxidative stress, bone demineralization, and cardiovascular diseases [34]. The target hazard quotient (THQ) and hazard risk index (HRI) are commonly used to evaluate the potential health risks associated with consuming PTE-contaminated food [35]. Elevated THQ values indicate a significant non-carcinogenic health risk, particularly for populations that consume *N. nucifera* rhizomes as a staple [36].

Despite increasing concerns about heavy metal contamination in aquatic environments, limited research has examined the bioaccumulation of PTEs in *N. nucifera*. Previous studies have reported PTE in various macrophytes, yet there is insufficient data on the extent of PTE uptake by *N. nucifera*, particularly in highly polluted freshwater ecosystems, i.e., Dal Lake. The contamination of water and sediment by anthropogenic activities raises concerns regarding metal accumulation in edible plant parts, which may pose health risks to consumers [37]. However, no study has specifically assessed the bioaccumulation and health risks associated with *N. nucifera* growing in the floating lake gardens of Dal Lake.

Thus, this study hypothesizes that *N. nucifera* accumulates PTEs from contaminated water and sediments, leading to potential health risks. The objectives of this study were: (1) to evaluate the physicochemical properties of water and sediment, as well as the bioaccumulation and translocation patterns of PTEs in different parts of *N. nucifera*; and (2) to assess the health risks associated with the consumption of *N. nucifera* rhizomes and seeds. The findings of this research provide valuable information for metal contamination in edible aquatic plants and contribute to environmental monitoring efforts, risk assessment frameworks, and potential remediation strategies for contaminated aquatic ecosystems.

2. Materials and Methods

2.1. Study Area

This study was conducted in Dal Lake, Srinagar, Kashmir (34.1106° N, 74.8683° E), a freshwater lake renowned for its scenic beauty and cultural significance (Figure 1). The lake, historically fed by the Jhelum River, has transformed into an urban water body due to increasing tourism and anthropogenic activities. It has a shoreline of approximately 15.5 km and covers an area of 18 km^2, with an average depth of 1.5 m and a maximum depth of 6 m. The lake is divided into four basins: Hazratbal, Bod Dal, Gagribal, and Nigeen. The present study was conducted between August and October 2024, coinciding with the growth and harvest period of *N. nucifera*, locally known as *Nadru*. *N. nucifera* cultivation is a traditional practice in the floating gardens (Rad in the local language) of Dal Lake, which serve as an important economic resource for local farmers. These gardens are established in shallow shore areas (1–2 m depth), where organic content-rich sediments provide suitable conditions for *N. nucifera* growth. The plant is cultivated for seeds (*Makhana*), rhizomes, and flowers, which hold nutritional, medicinal, and religious importance. Previously, the plant was genetically identified by Mehraj et al. [38] using the DNA barcoding method (MatK primer).

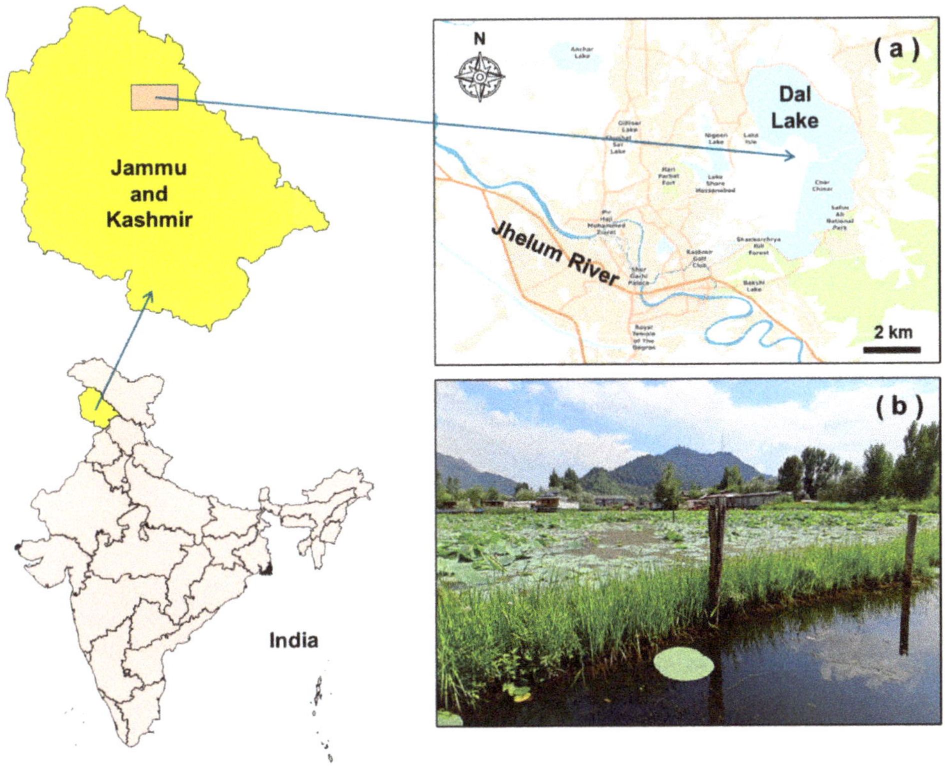

Figure 1. (**a**) Map of the Dal Lake located in Jammu and Kashmir state of India, and (**b**) floating lake gardens with *N. nucifera* (map created using Bhuvan 2D and Power-user).

2.2. Sampling Design and Sample Collection

A comprehensive sampling approach was employed to assess the uptake of potentially toxic elements in *Nelumbo nucifera* in Dal Lake, Srinagar. Samples were collected over three phases during the *N. nucifera* growth and harvest period, specifically from August to October 2024. The study area was divided into four distinct spatial zones—Hazratbal, Bod Dal, Gagribal, and Nigeen. In each zone, three samples were collected per month, resulting in a total of 12 samples per month. Over the three-month study period, 36 samples were collected separately for water, sediment, and plant tissues. In this, water samples were collected in 5 L capacity PVC containers, following standard procedures. These containers were pre-rinsed with distilled water before sample collection. Sediment samples were collected using a Van Veen grab sampler at a bottom depth of 5–10 cm [39]. The samples were transferred into pre-cleaned polyethylene bags, stored on ice, and transported to the laboratory for analysis. On the other hand, plant samples, including *N. nucifera* rhizomes, roots, petioles, leaves, and seeds, were carefully harvested by cutting using a sharp knife and stored in sterile zip-lock polybags to prevent microbial degradation and cross-contamination. All collected samples were transported under controlled conditions to the laboratory for further physicochemical and elemental analysis.

2.3. Analytical Methods

Water and sediment quality parameters were analyzed to assess the physicochemical characteristics of Dal Lake. The pH of water and sediment samples was measured using a digital pH meter to determine the acidity or alkalinity. Electrical conductivity (EC) and total dissolved solids (TDS) were analyzed using a multimeter. Dissolved organic matter (DOM) was quantified using a UV-Vis spectrophotometric method [40]. Biochemical oxygen demand (BOD_5) was determined using the Winkler method. Chemical oxygen demand (COD) was analyzed by the open reflux method, which estimates the oxidizable organic and inorganic matter in water. Total nitrogen (TN) was quantified using the Kjeldahl method [41]. Total phosphorus (TP) was estimated using a spectrophotometric method. PTE concentrations (Cd, Cu, Cr, Co, Fe, Mn, Ni, and Zn) in water and *N. nucifera* tissue were analyzed using Inductively Coupled Plasma Optical Emission Spectroscopy (ICP-OES). Water samples were acidified with nitric acid (HNO_3) and subjected to microwave digestion before ICP-OES analysis. Sediment and plant samples were oven-dried, ground into a fine powder, and digested using a mixture of HNO_3 and hydrogen peroxide (H_2O_2). The resulting digests from three matrices were filtered, diluted to 50 mL using 2% HNO_3, and analyzed using ICP-OES for quantification of PTE concentrations [42]. Method validation was performed by assessing recovery rates using Certified Reference Materials (CRM), with analytical accuracy within an acceptable range of 96–102%. The Limit of Detection (LOD) and Limit of Quantification (LOQ) were determined for each PTE using signal-to-noise ratios of 3:1 and 10:1, respectively.

2.4. Data Analysis

The bioconcentration factor (BCF) and translocation factor (TF) were calculated to assess the uptake and internal mobility of elements in *N. nucifera*. The BCF was determined as the ratio of PTE concentration in plant tissues to that in water. The TF was calculated as the ratio of PTE concentration in the aerial parts to that in the roots.

$$\mathrm{BAF} = \mathrm{PTEplant}/\mathrm{PTEwater} \quad (1)$$

$$\mathrm{TF} = \mathrm{PTEaerial}/\mathrm{PTEroot} \quad (2)$$

To evaluate potential health risks, dietary intake of metal (DIM), hazard risk index (HRI), and target hazard quotient (THQ) were computed for edible plant tissues, particularly rhizomes, using established risk assessment models [35]. In this, Equations (3) and (4) were used to compute HRI and DIM values, as given below:

$$\mathrm{HRI} = \frac{\mathrm{DIM}}{\mathrm{RfD}} \quad (3)$$

$$\mathrm{DIM} = \mathrm{SL} \times \frac{\mathrm{PTEc}}{\mathrm{Bw}} \quad (4)$$

where RfD, SL, PTEc, and Bw represent oral reference dose, serving of *N. nucifera* rhizome and seeds (dried weight), PTE concentration, and body weight of consumer (70 kg), respectively. Additionally, THQ [43] was used to evaluate the health risk of PTE accumulation in saffron as per the following model (Equation (5)):

$$\mathrm{THQ} = 10^{-3} \times \frac{\mathrm{Ef} \times \mathrm{Ed} \times \mathrm{Ir} \times \mathrm{PTEc}}{\mathrm{Bw} \times \mathrm{Cp} \times \mathrm{RfD}} \quad (5)$$

where 10^{-3} is the conversion factor, Ef is the exposure frequency, Ed is the exposure duration (365 days), Ir represents the saffron ingestion rate (0.30 kg/day), PTEc is the PTE concentration in the *N. nucifera* sample (mg/kg), Bw corresponds to the average body weight (70 kg), Cp is the consumption period (25550 days), and Rd is the reference dose in suggested by USEPA [44] terms of mg/kg/day (Cd: 1.0×10^{-3}; Cu: 4.0×10^{-2};

Cr: 5.0×10^{-3}; Co: 2.0×10^{-2}, Fe: 7.0×10^{-1}; Mn: 1.4×10^{-2}, Ni: 2.0×10^{-2}, and Zn: 3.0×10^{-1}). Further, the combined toxicity of PTE intake from *N. nucifera* was calculated as per Equation (6):

$$\sum \text{THQ} = \text{THQ}(\text{Cd} + \text{Cu} + \text{Cr} + \text{Co} + \text{Fe} + \text{Mn} + \text{Ni} + \text{Zn}) \quad (6)$$

To determine the contribution of each PTE in *N. nucifera* uptake, the Accumulation Nutrient Elements (ANEs) method was applied [45]. The ANEs model quantifies the uptake of elements by *N. nucifera* by assessing their accumulation in plant tissues. This model helps determine the contribution of each element to overall uptake, providing insights into bioaccumulation patterns. The percentage contribution of each PTE was computed using the following model (Equation (7)):

$$Y = \sum_{i=1}^{i} (Z\ :\ z) \quad (7)$$

where Y represents the total accumulation of participating PTE in mmol_c/kg. Z denotes the concentration of PTE (mg/kg), and z is its ion valency. The percentage contribution (X) of each PTE was then calculated using the formula (Equation (8)):

$$X = \left[\frac{(Z\ :\ z) \times 100}{Y}\right] \quad (8)$$

One-way analysis of variance (ANOVA), followed by Tukey's Honest Significant Difference (HSD) test, was conducted to compare PTE concentrations across different plant tissues (statistical significance at $p < 0.05$). All statistical analyses and graphical visualization were performed using MS Excel 365 (Microsoft Corp, Redmond, WA, USA) and OriginPro 2024b (OriginLab, Northampton, MA, USA).

3. Results and Discussion

3.1. Water and Sediment Quality of Dal Lake

As depicted in Table 1, the water quality parameters of Dal Lake exhibit significant variations from permissible limits set by CPCB and BIS. The average pH of lake water was recorded to be alkaline (8.12) while sediment showed an acidic reaction (6.30). The observed pH difference between lake water and sediment can be attributed to several environmental and geochemical factors. The alkaline nature of lake water is likely influenced by high biological productivity, photosynthetic activity, and carbonate buffering, which increase hydroxyl ion concentration [46]. In contrast, the sediment exhibits an acidic reaction due to organic matter decomposition, microbial activity, and the accumulation of metal sulfides, which release protons and lower pH. Additionally, limited mixing between sediment and overlying water prevents pH equilibration, contributing to this disparity [47]. The BOD of 15.72 mg/L exceeds the permissible limit of 3 mg/L, indicating high organic pollution and microbial activity, likely due to anthropogenic inputs. Similarly, COD is high (135.15 mg/L), which is the result of a considerable load of non-biodegradable organic pollutants. TN and TP concentrations in water are also high. PTE concentrations in water, particularly Cd (0.06 mg/L), Cr (0.30 mg/L), and Ni (0.45 mg/L), exceeded regulatory limits. On the other hand, the bottom sediment data showed substantial accumulation of PTEs, with Fe (1610.51 mg/kg), Mn (31.48 mg/kg), and Cr (29.72 mg/kg) exceeding background levels. Herein, Cd concentration (0.94 mg/kg) exceeds the Canadian Sediment Quality Guidelines (CSQG) threshold (0.68 mg/kg) [48] and national sediment quality guidelines (SQGs) [49]. Moreover, DOM in sediments (455.37 mg/kg) further suggests high organic contamination.

The observed EC in sediments (2.26 dS/m) was high compared to water (0.36 dS/m), which indicated ionic enrichment over time.

Table 1. Water and bottom sediment quality of floating gardens in Dal Lake, Srinagar, India.

Parameter	Water			Bottom Sediment		
	Value	CPCB Standard	BIS Standard	Value	CSQG Standard	USEPA Standard
pH	8.12 ± 0.05	6.50–9.00	6.50–8.50	6.30 ± 0.04	-	6.00−8.50
EC (dS/m)	0.36 ± 0.02	0.75	0.75	2.26 ± 0.51	-	4.00
TDS	242.85 ± 6.95 mg/L	500.00	500.00	-	-	-
DOM	22.95 ± 6.20 mg/L	-	-	455.37 ± 79.05 mg/kg	-	0.10−5.00%
BOD	15.72 ± 0.42 mg/L	3.00	3.00	-	-	-
COD	135.15 ± 12.85 mg/L	250.00	250.00	-	-	-
TN	9.89 ± 1.28 mg/L	-	-	581.04 ± 45.09 mg/kg	-	0.02−0.50%
TP	5.60 ± 0.46 mg/L	-	-	94.20 ± 8.71 mg/kg	-	200−1000
Cd	0.06 ± 0.03 mg/L	0.01	0.01	0.94 ± 0.12 mg/kg	0.68	4.98
Cu	3.40 ± 0.29 mg/L	1.50	1.50	13.24 ± 1.87 mg/kg	18.70	149.00
Cr	0.30 ± 0.07 mg/L	0.05	0.05	29.72 ± 3.90 mg/kg	52.30	111.00
Co	0.52 ± 0.10 mg/L	-	-	11.63 ± 0.42 mg/kg	-	-
Fe	2.95 ± 0.16 mg/L	0.30	0.30	1610.51 ± 249.03 mg/kg	2.00–5.00%	-
Mn	0.50 ± 0.07 mg/L	0.10	0.10	31.48 ± 3.72 mg/kg	-	460.00
Ni	0.45 ± 0.09 mg/L	0.02	0.02	4.10 ± 0.21 mg/kg	15.90	48.60
Zn	1.15 ± 0.03 mg/L	5.00	5.00	66.86 ± 8.32 mg/kg	124.00	459.00

Values are means followed by the standard deviation of 36 samples; CPCB: Central Pollution Control Board, India; BIS: Bureau of Indian Standards; CSQG: Canadian Sediment Quality Guidelines; USEPA: United States Environmental Protection Agency.

The observed water quality in Dal Lake indicated severe anthropogenic stress, primarily driven by unregulated sewage discharge, agricultural runoff, and industrial effluents [50]. The high BOD and COD levels indicate significant organic pollution, likely from untreated wastewater and decaying plant matter, which triggers microbial activity. Also, high nutrient levels, particularly TN and TP, indicated eutrophication possibilities [51], which could result in promoting algal blooms and oxygen depletion [52]. PTE contamination, particularly Cd, exceeding permissible limits, points to industrial and other commercial emissions, as well as leaching from urban waste. The acidic sediment pH and high EC also show that the lake has undergone long-term ionic enrichment and PTE accumulation, intensified by limited sediment removal. Previous studies on Dal Lake have shown significant natural and anthropogenic influences on its water and sediment chemistry. Water samples show a predominance of carbonate and silicate weathering, with lower pH and higher total dissolved solids in some areas due to sewage inputs [28]. PTE analysis also showed moderate enrichment of Cr, Ni, Cu, Zn, Pb, Fe, and Mn in sediments, with pollution levels increasing towards the central parts of the lake. Higher total organic carbon and nitrogen contents suggest eutrophic conditions in the lake basin [53]. In a study, Ahamad et al. [27] compared Dal and Nigeen lakes, showing higher levels of BOD, EC, COD, and PO_4^{3-} in Dal Lake. Similarly, Kumar et al. [24] showed a significant decline in water quality over the past 40 years due to anthropogenic pressures, with decreased transparency and dissolved oxygen, and increased phosphates, nitrates, and chlorides.

3.2. PTE Concentrations in *N. nucifera* Tissues

As shown in Table 2, the distribution of PTEs in *N. nucifera* tissues varied significantly, with roots exhibiting the highest accumulation across most elements. Specifically, Cd was concentrated in roots (0.45 ± 0.05 mg/kg), significantly higher than in rhizomes (0.15 ± 0.03 mg/kg) and petioles (0.13 ± 0.02 mg/kg), while leaves and seeds contained the lowest concentrations. Similarly, Cu followed a similar trend, with roots accumulating the highest levels (9.50 ± 0.22 mg/kg), whereas seeds had the lowest (1.80 ± 0.05 mg/kg). Also, Cr and Co showed the highest accumulation in roots (3.60 ± 0.10 and 2.85 ± 0.08 mg/kg, respectively), with significant reductions in aerial parts. However, Fe was most abundant in roots (280.57 ± 18 mg/kg), followed by leaves (173.96 ± 8 mg/kg) and petioles (150.76 ± 10 mg/kg). Likewise, Mn and Ni concentrations were highest in roots, with significant declines in rhizomes, petioles, and leaves. Similarly, Zn exhibited the maximum accumulation in roots (40.30 ± 1.20 mg/kg), while seeds had the lowest levels (7.50 ± 0.25 mg/kg).

Table 2. The concentration (mg/kg dwt.) of PTEs in different plant tissues of *N. nucifera* grown in the floating lake gardens of Dal Lake.

PTE	*N. nucifera* Tissues				
	Rhizome	Roots	Petiole	Leaves	Seeds
Cd	0.15 ± 0.03 b	0.45 ± 0.05 a	0.13 ± 0.02 b	0.04 ± 0.01 c	0.08 ± 0.01 c
Cu	6.25 ± 0.15 b	9.50 ± 0.22 a	4.80 ± 0.10 c	3.20 ± 0.08 d	1.80 ± 0.05 e
Cr	2.10 ± 0.07 b	3.60 ± 0.10 a	1.75 ± 0.05 c	1.10 ± 0.03 d	0.65 ± 0.02 e
Co	1.75 ± 0.06 b	2.85 ± 0.08 a	1.30 ± 0.04 c	0.90 ± 0.02 d	0.50 ± 0.01 e
Fe	120.06 ± 12 c	280.57 ± 18 a	150.76 ± 10 b	173.96 ± 8 b	105.40 ± 5 c
Mn	8.50 ± 0.20 b	12.75 ± 0.35 a	6.40 ± 0.15 c	4.25 ± 0.10 d	2.10 ± 0.05 e
Ni	2.60 ± 0.08 b	4.30 ± 0.12 a	2.00 ± 0.06 c	1.40 ± 0.04 d	0.75 ± 0.02 e
Zn	25.50 ± 0.80 b	40.30 ± 1.20 a	18.40 ± 0.55 c	12.60 ± 0.40 d	7.50 ± 0.25 e

Values are means followed by the standard deviation of 36 samples; different letters within each row indicate significant differences among plant tissues ($p < 0.05$) based on Tukey's HSD.

The uptake of PTEs by *N. nucifera* from lake water and sediment is affected by several factors such as root system, metal-binding capacity, and physiological adaptations. Roots act as primary sites for metal absorption due to their direct contact with sediment, where metals exist in bioavailable forms [54]. In particular, Cd is taken up due to its similarity to essential elements like Ca^{2+}, allowing it to enter through Ca^{2+} transport channels [55]. Meanwhile, Cu is essential for enzymatic functions and is actively absorbed, but can accumulate excessively under contamination [56]. Similarly, Cr, particularly Cr(III), is absorbed due to its resemblance to essential micronutrients like Fe, while Co is taken up due to its role in nitrogen metabolism [57,58]. Fe is essential for chlorophyll synthesis, leading to high accumulation as evident in this study [59]. Also, Mn assists in photosynthesis and enzyme activation, Ni is required in trace amounts for urease activity, and Zn is absorbed as a cofactor for numerous enzymes [60]. Previously, Abd Rasid et al. [61] investigated the potential of *N. nucifera* in treating surface water and found that it can significantly reduce pollutant load while accumulating high levels of different mineral elements. Similarly, Liu et al. [62] also reported that *N. nucifera* could accumulate high levels of Cd in its tissues and clean the polluted waters. However, high Cd levels in plants could bring negative impacts on plants as well as consumers. In another study, Painuly et al. [63] also determined the As(III) accumulation capacity of *N. nucifera*. [64] found that *N. nucifera* has the potential to accumulate nine PTEs (Zn, Cu, Pb, Ni, Mn, Hg, Cr, Cd, and As) from irrigation lakes in Debarawewa and Galewela provinces in Sri Lanka.

3.3. Bioaccumulation and Translocation Factors

The BCF indicates the efficiency of *N. nucifera* in accumulating PTEs from its water body to its vegetative tissues (Table 3). In this, roots exhibited the highest BCF values for all PTEs, particularly Fe (95.11), Cr (12.00), and Mn (25.50), indicating their strong metal-binding capacity. Nevertheless, rhizomes also showed high accumulation, especially for Fe (40.70) and Zn (22.17). For aerial parts such as petioles and leaves, lower BCFs were recorded, exhibiting limited translocation from roots. Seeds consistently exhibited the lowest BCF values, which might be due to restricted PTE mobility. The high root BCF values showed the phytostabilization potential, where PTEs are sequestered in belowground tissues, minimizing translocation to edible parts. On the other hand, TF values (Table 4) showed the translocation efficiency of PTEs from roots to different tissues of *N. nucifera*. The highest TF was observed for Fe (0.62) from roots to leaves, which might be due to its active translocation via xylem transport for metabolic functions. Other PTEs exhibited lower translocation efficiencies, with Cd (0.09) showing the least mobility to leaves, suggesting strong retention in roots. Mn (0.67) and Cu (0.66) had relatively higher TF values from roots to rhizomes, indicating their moderate mobility. However, PTEs showed limited translocation to seeds (TF < 0.40). As given in Table 5, the results of ANE modeling showed a proportional contribution of each PTE in *N. nucifera* tissues. In this, Fe exhibited the highest accumulation across all tissues, accounting for 80.95–92.92% of total PTE uptake, indicating its requirement for plant metabolism and its strong affinity for root and shoot tissues. However, Zn also demonstrated a significant accumulation (6.10–9.79%), particularly in roots and rhizomes. Mn contributed 1.25–3.88%, while uptake of non-essential PTEs such as Cd, Cr, and Ni remained below 1.11%, which might be due to restricted translocation and possible detoxification mechanisms. In this, root tissues consistently accumulated the highest PTE concentrations, reinforcing their role as primary sites for PTE sequestration. The low participation of Cd (0.01–0.05%) indicates its minimal uptake, likely due to its toxicity. Overall, the results indicate that *N. nucifera* primarily stabilizes PTEs in below-ground tissues, minimizing translocation to aerial parts.

Table 3. Bioconcentration factors (BCFs) of PTE uptake by *N. nucifera* plant tissues.

BCF	***N. nucifera* Tissues**				
	Rhizome	**Roots**	**Petiole**	**Leaves**	**Seeds**
Cd	2.50	7.50	2.17	0.67	1.33
Cu	1.84	2.79	1.41	0.94	0.53
Cr	7.00	12.00	5.83	3.67	2.17
Co	3.37	5.48	2.50	1.73	0.96
Fe	40.70	95.11	51.11	58.97	35.73
Mn	17.00	25.50	12.80	8.50	4.20
Ni	5.78	9.56	4.44	3.11	1.67
Zn	22.17	35.04	16.00	10.96	6.52

Values are means followed by the standard deviation of 36 samples.

Table 4. Translocation factors (TFs) of PTE uptake by *N. nucifera* plant tissues.

TF	***N. nucifera* Tissues**			
	Root → Rhizome	**Root → Petiole**	**Root → Leaves**	**Root → Seed**
Cd	0.33	0.29	0.09	0.18
Cu	0.66	0.51	0.34	0.19
Cr	0.58	0.49	0.31	0.18
Co	0.61	0.46	0.32	0.18
Fe	0.43	0.54	0.62	0.38
Mn	0.67	0.50	0.33	0.16
Ni	0.60	0.47	0.33	0.17
Zn	0.63	0.46	0.31	0.19

Values are means followed by the standard deviation of 36 samples.

Table 5. Accumulation of nutrient elements (ANEs) modeling for PTE uptake by *N. nucifera* plant tissues.

Element	*N. nucifera* Tissues									
	Rhizome		Roots		Petiole		Leaves		Seeds	
	X	*P* (%)	*X*	*P* (%)	*X*	*P* (%)	*X*	*P* (%)	*X*	*P* (%)
∑PTE ($mmol_c$/Kg)	796.73	100.00	1745.79	100.00	922.77	100.00	1010.73	100.00	609.36	100.00
Cd	0.27	0.03	0.80	0.05	0.23	0.03	0.07	0.01	0.14	0.02
Cu	19.67	2.47	29.90	1.71	15.11	1.64	10.07	1.00	5.67	0.93
Cr	8.08	1.01	13.85	0.79	6.73	0.73	4.23	0.42	2.50	0.41
Co	5.94	0.75	9.67	0.55	4.41	0.48	3.05	0.30	1.70	0.28
Fe	644.96	80.95	1507.23	86.33	809.88	87.77	934.52	92.46	566.21	92.92
Mn	30.94	3.88	46.42	2.66	23.30	2.52	15.47	1.53	7.64	1.25
Ni	8.86	1.11	14.65	0.84	6.82	0.74	4.77	0.47	2.56	0.42
Zn	78.01	9.79	123.28	7.06	56.29	6.10	38.54	3.81	22.94	3.77

Values are means followed by the standard deviation of 36 samples; *X*: total accumulation of the respective PTEs in *N. nucifera* tissues, expressed in $mmol_c$/kg; *P* (%): percentage contribution of the respective PTE to the total PTE accumulation in each tissue. WHO permissible limits in vegetables: Cd: 0.02, Cu: 10.00, Cr: 1.30, Ni: 10.00, Zn: 0.60, Co, Fe, and Mn: not defined.

Several studies have investigated bioaccumulation and translocation factors in lake plants and modeled PTE accumulation. Galal and Farahat [65] examined *Pistia stratiotes* in Lake Mariut, finding bioaccumulation factors greater than one for most PTEs, except Cu, while translocation factors were also less than 1, suggesting its suitability for rhizofiltration. Bai et al. [66] evaluated the bioaccumulation potential of four aquatic plants in Taihu Lake, revealing that submerged plants, especially their stems, showed a closer relationship with PTEs in water and sediment compared to floating-leaf plants. Also, Skorbiłowicz et al. [67] used *Phragmites australis* as a bioindicator in the Bug River catchment, demonstrating that roots accumulated the highest levels of potentially toxic elements, making them necessary for monitoring PTE concentrations. In another study, Parzych et al. [45] applied ANE modeling to assess the capacity of *Salix viminalis* (willow) leaves and bark to accumulate PTEs from contaminated environments. They quantified the concentration of PTEs such as Cd, Pb, and Zn to assess the plant's potential for phytoremediation and environmental monitoring.

3.4. Health Risk Assessment of PTEs

As shown in Table 6, the health risk assessment of PTE in *N. nucifera* edible tissues (rhizome and seeds only) was evaluated using HRI, DIM, and THQ indices. The results indicate that Cr exhibited the highest HRI in both rhizomes (1.8000) and seeds (0.5571), suggesting potential non-carcinogenic health risks upon prolonged consumption. Other PTEs, including Cd, Cu, and Fe, showed moderate HRI values, while Co, Mn, and Zn exhibited the lowest. The DIM values followed a similar trend, with Fe showing the highest intake, particularly in rhizomes (0.5143 mg/kg/day), while Cd had the lowest across both tissues. The THQ values for all PTEs remained below 1, indicating no immediate non-carcinogenic risk. The cumulative THQ (∑THQ) was 0.0054 for rhizomes and 0.0021 for seeds, further confirming low health risks. However, chronic exposure to Cr and Cd, given their relatively higher HRI, necessitates further monitoring. PTE contamination in edible *N. nucifera* tissues poses health risks to consumers.

The accumulation of PTEs in *N. nucifera* has ecological and health implications, as these elements can enter the food chain and affect aquatic ecosystems. Specifically, Cr is carcinogenic and induces oxidative stress, leading to cellular damage, which may impact aquatic organisms and plant health [68]. Plants primarily take up chromium in the hexavalent form [Cr(VI)] through root absorption, as it is more soluble and mobile in water compared to trivalent chromium [Cr(III)] [69]. However, Cr(III) can also be absorbed to a lesser extent if present in a bioavailable chelated form. In this, Cr(VI) is known to be more toxic than Cr(III) [70]. If contaminated *N. nucifera* is consumed by humans, Cd could accumulate in the kidneys and liver, causing renal dysfunction and bone demineralization, raising concerns about its bioaccumulation in edible plant parts [71]. While Cu is essential for enzymatic functions, excessive intake results in gastrointestinal distress and liver toxicity, posing risks to both plants and herbivorous consumers [72]. Fe is crucial for hemoglobin synthesis, but elevated levels contribute to oxidative stress and organ damage, potentially affecting plant physiology and aquatic biota [73]. Mn supports metabolism, yet excessive amounts impair neurological function, which may influence the health of organisms relying on contaminated water sources [74]. Ni exposure is linked to respiratory and cardiovascular issues, emphasizing risks for both aquatic and terrestrial ecosystems [75]. Co, essential for vitamin B12 synthesis, can be cardiotoxic at high concentrations, necessitating careful monitoring in sediment and water. Excess Zn disrupts immune function and nutrient absorption, potentially altering plant growth and ecosystem balance [76,77]. Understanding these effects is crucial for assessing environmental contamination, managing sediment quality, and ensuring the safe use of *N. nucifera* in food and medicine.

Table 6. Health risk assessment of PTEs in edible parts of *N. nucifera* plants grown in floating lake garden.

Element	HRI		DIM		THQ	
	Rhizome	Seeds	Rhizome	Seeds	Rhizome	Seeds
Cd	0.6429	0.3429	0.0006	0.0003	0.0006	0.0003
Cu	0.6696	0.1929	0.0268	0.0077	0.0007	0.0002
Cr	1.8000	0.5571	0.0090	0.0028	0.0018	0.0006
Co	0.3750	0.1071	0.0075	0.0021	0.0004	0.0001
Fe	0.7347	0.6429	0.5143	0.4500	0.0007	0.0006
Mn	0.2602	0.0643	0.0364	0.0090	0.0003	0.0001
Ni	0.5571	0.1607	0.0111	0.0032	0.0006	0.0002
Zn	0.3643	0.1071	0.1093	0.0321	0.0004	0.0001
∑THQ	-	-	-	-	0.0054	0.0021

Values are means followed by the standard deviation of 36 samples; Health Risk Index: HRI < 1 (safe), HRI ≥ 1 (potential risk); Daily Intake of Metal: Compared against tolerable daily intake limits of WHO/USEPA [78] (Cd: 1.00 µg/kg, Cu: 0.50 mg/kg, Cr: 25.00 µg/kg, Fe: 0.8 mg/kg, Mn: 0.06 mg/kg, Ni: 0.005 mg/kg, Zn: 1.00 mg/kg, Co: not defined); target hazard quotient: THQ < 1 (no risk), THQ ≥ 1 (potential health risk); ∑THQ (Cumulative Risk): ∑THQ < 1 (safe), ∑THQ ≥ 1 (potential non-carcinogenic risk).

Previously, Ologundudu et al. [79] assessed the PTE bioaccumulation potential of *Corchorus olitorius* (L.) and *Amaranthus hybridus* (L.) collected from a polluted dumpsite and found increased values of DIM and HRI. Obasi et al. [80] studied the health risks of PTEs (Hg, Cd, Mn, Pb, Cu, Co, As, Cr, Zn, and Mo) and found that Hg, Cd, and Pb had HRI and THQ > 1. In the Jammu and Kashmir region of India, Abou Fayssal et al. [35] studied the health risk of PTE uptake by saffron (*Crocus sativus* L.) cultivated in soils irrigated with domestic wastewater and Sarbal Lake water. They found that crops irrigated with domestic and Sarbal Lake water both had higher values of HRI, DIM, and THQ compared to those irrigated with borewell supplies. Therefore, these studies also corroborate the findings of the present study on the potential health risks associated with edible crop contamination grown in lakes.

4. Conclusions

The results of the present study indicated a significant accumulation of PTEs in *N. nucifera* cultivated in Dal Lake. The study revealed significant PTE contamination, with high sediment concentrations of Fe (1610.51 mg/kg), Mn (31.48 mg/kg), and Cr (29.72 mg/kg), preferential accumulation in roots (Fe uptake: 95.11), low translocation to edible parts, and Cr posing the highest non-carcinogenic risk (HRI = 1.8000 in rhizomes), while the cumulative THQ remained below 1, indicating no immediate risk but potential concerns with long-term exposure. In this, PTEs were primarily sequestered in root tissues, thereby reducing their translocation to edible parts. While the THQ values indicate no immediate health risk, the high HRI for Cr necessitates continued monitoring, especially given the chronic toxicity associated with PTE exposure. The findings are useful for policymakers and environmental agencies aiming to regulate lake water quality and mitigate contamination sources. However, this study has limitations, including the absence of long-term exposure assessments and potential variability in PTEs due to seasonal changes. Future research should focus on remediation strategies, such as phytoremediation enhancements, and evaluate the impact of varying environmental conditions on PTE bioaccumulation. Further studies are also needed to explore the genetic and physiological responses of *N. nucifera* to PTE stress, which may aid in the development of bioengineered aquatic plants for improved phytoremediation efficacy.

Author Contributions: Conceptualization, M.E., P.K. and I.Š.; Data curation, F.S.M., I.M., S.E.-N., A.E.-D.O., B.M. and Ž.A.; Formal analysis, M.G. and P.K.; Funding acquisition, M.E.; Investigation, M.G. and P.K.; Methodology, M.E., A.E.-D.O., M.G., P.K., B.M. and I.Š.; Project administration, M.E.; Resources, P.K.; Software, F.S.M., I.M., S.E.-N., A.E.-D.O., B.M. and Ž.A.; Supervision, M.E. and I.Š.; Validation, M.E., F.S.M., I.M., S.E.-N., A.E.-D.O., P.K., B.M., Ž.A. and I.Š.; Visualization, F.S.M., I.M., S.E.-N., A.E.-D.O., M.G., B.M., Ž.A. and I.Š.; Writing—original draft, M.G., P.K. and I.Š.; Writing—review and editing, M.E., F.S.M., I.M., S.E.-N., A.E.-D.O., B.M. and Ž.A. All authors have read and agreed to the published version of the manuscript.

Funding: The Deanship of Research and Graduate Studies at King Khalid University funded this study through the Large Research Project under grant number RGP2/118/45.

Institutional Review Board Statement: Not applicable.

Informed Consent Statement: Not applicable.

Data Availability Statement: The original contributions presented in this study are included in the article. Further inquiries can be directed to the corresponding author.

Acknowledgments: The authors extend their appreciation to the Deanship of Research and Graduate Studies at King Khalid University for funding this study through the Large Research Project under Grant number RGP2/118/45. All individuals included in this section have consented to the acknowledgement.

Conflicts of Interest: The authors declare no conflicts of interest.

References

1. Yang, H.; He, S.; Feng, Q.; Liu, Z.; Xia, S.; Zhou, Q.; Wu, Z.; Zhang, Y. Lotus (*Nelumbo nucifera*): A Multidisciplinary Review of Its Cultural, Ecological, and Nutraceutical Significance. *Bioresour. Bioprocess.* **2024**, *11*, 18. [CrossRef] [PubMed]
2. Luo, M.; Jia, H.; Li, Q.; Meng, X.; Ferguson, D.K.; Liu, P.; Han, Z.; Wang, J.; Quan, C. Middle Miocene Lotus (Nelumbonaceae, *Nelumbo*) from the Qaidam Basin, Northern Tibet Plateau. *Biology* **2022**, *11*, 1261. [CrossRef] [PubMed]
3. Chen, L.; Zhu, G.; Lin, X.; Li, R.; Lu, S.; Jiao, Y.; Wang, Q. The Complexity of Moisture Sources Affects the Altitude Effect of Stable Isotopes of Precipitation in Inland Mountainous Regions. *Water Resour. Res.* **2024**, *60*, e2023WR036084. [CrossRef]
4. Wang, Q.; Liu, Y.; Zhu, G.; Lu, S.; Chen, L.; Jiao, Y.; Li, W.; Li, W.; Wang, Y. Regional differences in the effects of atmospheric moisture residence time on precipitation isotopes over Eurasia. *Atmos. Res.* **2025**, *314*, 107813. [CrossRef]
5. Lin, Z.; Zhang, C.; Cao, D.; Damaris, R.N.; Yang, P. The Latest Studies on Lotus (*Nelumbo nucifera*)—An Emerging Horticultural Model Plant. *Int. J. Mol. Sci.* **2019**, *20*, 3680. [CrossRef]
6. Jafari, N.; Trivedy, R.K. Ecological Integrity of Wetlands, Their Functions and Sustainable Use with a Case Study of Anjali Wetland, Iran. *Ecol. Environ. Conserv.* **2009**, *15*, 191–199.

7. Sharip, Z.; Zaki, A.T.A.; Zakaria, S. Flooding Effects on the Population Dynamics of Cabomba Furcata and *Nelumbo nucifera* in a Shallow Floodplain Wetland. *Wetlands* **2014**, *34*, 713–723. [CrossRef]
8. Gupta, P.; Sharma, V.K.; Sharma, S. *Healing Traditions of the Northwestern Himalayas*; Springer: Mumbai, India, 2014; ISBN 978-81-322-1924-8.
9. Paudel, K.R.; Panth, N. Phytochemical Profile and Biological Activity of *Nelumbo nucifera*. *Evid. Based Complement. Altern. Med.* **2015**, *2015*, 789124. [CrossRef]
10. Chen, G.; Zhu, M.; Guo, M. Research Advances in Traditional and Modern Use of *Nelumbo nucifera*: Phytochemicals, Health Promoting Activities and Beyond. *Crit. Rev. Food Sci. Nutr.* **2019**, *59*, S189–S209. [CrossRef]
11. McDonald, A. A Botanical Perspective on the Identity of Soma (*Nelumbo nucifera* Gaertn.) Based on Scriptural and Iconographic Records. *Econ. Bot.* **2004**, *58*, S147–S173. [CrossRef]
12. Lan, T.; Hu, Y.; Cheng, L.; Chen, L.; Guan, X.; Yang, Y.; Pan, J. Floods and diarrheal morbidity: Evidence on the relationship, effect modifiers, and attributable risk from Sichuan Province, China. *J. Glob. Health* **2022**, *12*, 11007. [CrossRef] [PubMed]
13. Gowthami, R.; Sharma, N.; Pandey, R.; Agrawal, A. A Model for Integrated Approach to Germplasm Conservation of Asian Lotus (*Nelumbo nucifera* Gaertn.). *Genet. Resour. Crops Evol.* **2021**, *68*, 1269–1282. [CrossRef]
14. Chen, C.; Wang, R.; Chen, M.; Zhao, J.; Li, H.; Ignatieva, M.; Zhou, W. The post-effects of landscape practices on spontaneous plants in urban parks. *Urban For. Urban Green.* **2025**, *107*, 128744. [CrossRef]
15. Zhang, H.; Chen, G.; Zhang, Y.; Yang, M.; Chen, J.; Guo, M. Potential Hypoglycemic, Hypolipidemic, and Anti-Inflammatory Bioactive Components in *Nelumbo nucifera* Leaves Explored by Bioaffinity Ultrafiltration with Multiple Targets. *Food Chem.* **2022**, *375*, 131856. [CrossRef]
16. Goel, A.; Sharma, S.C.; Sharga, A.N. The Conservation of the Diversity of *Nelumbo* (Lotus) at the National Botanical Research Institute, Lucknow, India. *Bot. Gard. Conserv. News* **2001**, *3*, 52–54.
17. Sen, C.T.; Bhattacharyya, S.; Saberi, H. *The Bloomsbury Handbook of Indian Cuisine*; Bloomsbury: London, UK, 2023.
18. Chaudhuri, A.; Aqil, M.; Qadir, A. Herbal Cosmeceuticals: New Opportunities in Cosmetology. *Trends Phytochem. Res.* **2020**, *4*, 117–142.
19. Chen, W.J.; Jan, J.F.; Chung, C.H.; Liaw, S.C. Evaluating the Economic Viability of Agro-Ecotourism as a Nature-Based Solution for a Climate Adaptation Strategy: A Case Study of Yuanshan Township, Taiwan. *Sustainability* **2024**, *16*, 8267. [CrossRef]
20. Casimir, M.J. *Floating Economies*; Berghahn Books: New York, NY, USA, 2022.
21. Oo, M.T.; Aung, Z.W.; Puzzo, C. The Floating Garden Agricultural System of the Inle Lake (Myanmar) as an Example of Equilibrium between Food Production and Biodiversity Maintenance. *Biodivers. Conserv.* **2022**, *31*, 2435–2452. [CrossRef]
22. Irfanullah, H.M.; Adrika, A.; Ghani, A.; Khan, Z.A.; Rashid, M.A. Introduction of Floating Gardening in the North-Eastern Wetlands of Bangladesh for Nutritional Security and Sustainable Livelihood. *Renew. Agric. Food Syst.* **2008**, *23*, 89–96. [CrossRef]
23. Sheergojri, I.A.; Rashid, I.; Aneaus, S.; Rashid, I.; Qureshi, A.A.; Rehman, I.U. Enhancing the Social-Ecological Resilience of an Urban Lake for Sustainable Management. *Environ. Dev. Sustain.* **2023**, *27*, 8085–8110. [CrossRef]
24. Kumar, R.; Parvaze, S.; Huda, M.B.; Allaie, S.P. The Changing Water Quality of Lakes—A Case Study of Dal Lake, Kashmir Valley. *Environ. Monit. Assess.* **2022**, *194*, 228. [CrossRef] [PubMed]
25. Amin, A.; Fazal, S.; Mujtaba, A.; Singh, S.K. Effects of Land Transformation on Water Quality of Dal Lake, Srinagar, India. *J. Indian. Soc. Remote Sens.* **2014**, *42*, 119–128. [CrossRef]
26. Feng, X.; Wang, Z.; Wu, X.; Huang, S.; Li, J.; Lai, C.; Zeng, Z.; Lin, G. Tracking 3D Drought Events Across Global River Basins: Climatology, Spatial Footprint, and Temporal Changes. *Geophys. Res. Lett.* **2025**, *52*, e2024GL111442. [CrossRef]
27. Ahamad, F.; Sharma, A.K.; Tyagi, S.K. A Study on Comparative Assessment of Water Quality of Dal and Nigeen Lakes of Jammu and Kashmir, India. *AgroEnvironmental Sustain.* **2023**, *1*, 48–56. [CrossRef]
28. Jeelani, G.; Shah, A.Q. Geochemical Characteristics of Water and Sediment from the Dal Lake, Kashmir Himalaya: Constraints on Weathering and Anthropogenic Activity. *Environ. Geol.* **2006**, *50*, 12–23. [CrossRef]
29. Wen, J.; Zhang, J.; Zhang, H.; Zhang, N.; Lei, R.; Deng, Y.; Cheng, Q.; Li, H.; Luo, P. Large-scale genome-wide association studies reveal the genetic causal etiology between air pollutants and autoimmune diseases. *J. Transl. Med.* **2024**, *22*, 392. [CrossRef]
30. Taher, M.A.; Zouidi, F.; Kumar, P.; Abou Fayssal, S.; Adelodun, B.; Goala, M.; Kumar, V.; Andabaka, Ž.; Širić, I.; Eid, E.M. Impact of Irrigation with Contaminated Water on Heavy Metal Bioaccumulation in Water Chestnut (*Trapa natans* L.). *Horticulturae* **2023**, *9*, 190. [CrossRef]
31. Rawat, V.; Singh, A.K. Environmental and Ecological Importance of Indian Aquatic Macrophytes. In *Aquatic Macrophytes: Ecology, Functions and Services*; Springer: Singapore, 2023; pp. 71–83, ISBN 9789819938223.
32. AL-Huqail, A.A.; Kumar, P.; Eid, E.M.; Taher, M.A.; Kumar, P.; Adelodun, B.; Andabaka, Ž.; Mioč, B.; Držaić, V.; Bachheti, A.; et al. Phytoremediation of Composite Industrial Effluent Using Sacred Lotus (*Nelumbo nucifera* Gaertn): A Lab-Scale Experimental Investigation. *Sustainability* **2022**, *14*, 9500. [CrossRef]

33. Rizwan, M.; Ali, S.; Qayyum, M.F.; Ok, Y.S.; Adrees, M.; Ibrahim, M.; Zia-ur-Rehman, M.; Farid, M.; Abbas, F. Effect of Metal and Metal Oxide Nanoparticles on Growth and Physiology of Globally Important Food Crops: A Critical Review. *J. Hazard. Mater.* **2017**, *322*, 2–16. [CrossRef]
34. Ciobanu, C.; Slencu, B.G.; Cuciureanu, R. Estimation of Dietary Intake of Cadmium and Lead through Food Consumption. *Rev. Med. Chir. Soc. Med. Nat. Iasi* **2012**, *116*, 617–623.
35. Abou Fayssal, S.; Kumar, P.; Popescu, S.M.; ud din Khanday, M.; Sardar, H.; Ahmad, R.; Gupta, D.; Kumar Gaur, S.; Alharby, H.F.; Al-Ghamdi, A.G. Health Risk Assessment of Heavy Metals in Saffron (*Crocus sativus* L.) Cultivated in Domestic Wastewater and Lake Water Irrigated Soils. *Heliyon* **2024**, *10*, e27138. [CrossRef]
36. Ain, S.N.U.; Abbasi, A.M.; Ajab, H.; Faridullah; Khan, S.; Yaqub, A. Assessment of Arsenic in *Mangifera indica* (Mango) Contaminated by Artificial Ripening Agent: Target Hazard Quotient (THQ), Health Risk Index (HRI) and Estimated Daily Intake (EDI). *Food Chem. Adv.* **2023**, *3*, 100468. [CrossRef]
37. Wei, S.; Liu, X.; Tao, Y.; Wang, X.; Lin, Z.; Zhang, Y.; Hu, Q.; Wang, L.; Qu, J.; Zhang, Y. Strategy for Enhanced Soil Lead Passivation and Mitigating Lead Toxicity to Plants by Biochar-Based Microbial Agents. *J. Hazard. Mater.* **2025**, *489*, 137512. [CrossRef] [PubMed]
38. Mehraj, S.; Parihar, T.J.; Murtaza, D.; Hurrah, A.A.; Wani, I.A.; Lone, F.A.; Mufti, S.; Zargar, S.M.; Khan, I.; Sheikh, P.A.; et al. Macrophytes in Northern Himalayan Dal Lake of Kashmir Valley Identified through DNA Barcoding Shows High Antioxidant Potential. *Ecol. Genet. Genom.* **2023**, *27*, 100162. [CrossRef]
39. Van Hoey, G.; Birchenough, S.N.R.; Hostens, K. Estimating the Biological Value of Soft-Bottom Sediments with Sediment Profile Imaging and Grab Sampling. *J. Sea Res.* **2014**, *86*, 1–12. [CrossRef]
40. Li, P.; Hur, J. Utilization of UV-Vis Spectroscopy and Related Data Analyses for Dissolved Organic Matter (DOM) Studies: A Review. *Crit. Rev. Environ. Sci. Technol.* **2017**, *47*, 131–154. [CrossRef]
41. Bremner, J.M. Determination of Nitrogen in Soil by the Kjeldahl Method. *J. Agric. Sci.* **1960**, *55*, 11–33. [CrossRef]
42. Kumar, P.; Kumar, V.; Eid, E.M.; AL-Huqail, A.A.; Adelodun, B.; Abou Fayssal, S.; Goala, M.; Arya, A.K.; Bachheti, A.; Andabaka, Ž.; et al. Spatial Assessment of Potentially Toxic Elements (PTE) Concentration in *Agaricus bisporus* Mushroom Collected from Local Vegetable Markets of Uttarakhand State, India. *J. Fungi* **2022**, *8*, 452. [CrossRef]
43. Mahabadi, M. Assessment of Heavy Metals Contamination and the Risk of Target Hazard Quotient in Some Vegetables in Isfahan. *Pollution* **2020**, *6*, 69–78.
44. U.S. Environmental Protection Agency. *A Review of the Reference Dose and Reference Concentration Processes*; EPA/630/P-02/002F; U.S. Environmental Protection Agency: Washington, DC, USA, 2002. Available online: https://www.epa.gov/scientific-leadership/review-reference-dose-and-reference-concentration-processes (accessed on 10 January 2025).
45. Parzych, A.E.; Cymer, M.; Jonczak, J.; Szymczyk, S. The Ability of Leaves and Rhizomes of Aquatic Plants to Accumulate Macro- and Micronutrients. *J. Ecol. Eng.* **2015**, *16*, 198–205. [CrossRef]
46. Brosset, C. Factors Influencing PH in Lake Water. *Water Air Soil Pollut.* **1979**, *11*, 57–61. [CrossRef]
47. Herlihy, A.T.; Mills, A.L. The PH Regime of Sediments Underlying Acidified Waters. *Biogeochemistry* **1986**, *2*, 95–99. [CrossRef]
48. Canadian Sediment Quality Guidelines. Chemical-Specific Sediment Quality Guidelines: Canada. National Guidelines and Standards Office, Issuing Body; 2003. Available online: https://publications.gc.ca/site/archivee-archived.html?url=https://publications.gc.ca/collections/collection_2024/eccc/En13-11-2-2003-eng.pdf (accessed on 10 January 2025).
49. Bramha, S.N.; Mohanty, A.K.; Satpathy, K.K.; Kanagasabapathy, K.V.; Panigrahi, S.; Samantara, M.K.; Prasad, M.V.R. Heavy Metal Content in the Beach Sediment with Respect to Contamination Levels and Sediment Quality Guidelines: A Study at Kalpakkam Coast, Southeast Coast of India. *Environ. Earth Sci.* **2014**, *72*, 4463–4472. [CrossRef]
50. Parray, S.Y.; Koul, B.; Shah, M.P. Comparative Assessment of Dominant Macrophytes and Limnological Parameters of Dal Lake and Chatlam Wetlands in the Union Territory of Jammu & Kashmir, India. *Environ. Technol. Innov.* **2021**, *24*, 101978. [CrossRef]
51. Qu, J.; Wang, S.; Li, Z.; Wei, S.; Bi, F.; Yan, S.; Yu, H.; Wang, L.; Zhang, Y. Highly Efficient Recovery of Phosphate from Water Using Cerium Carbonate Hydroxide-Decorated Magnetic Biochar: A Slow-Release Phosphate Fertilizer for Agricultural Reuse. *ACS EST Engg.* **2024**, *12*, 3045–3056. [CrossRef]
52. Zaffar, R.; Nazir, R.; Rather, M.A.; Dar, R. Biofilm Formation and EPS Production Enhances the Bioremediation Potential of Pseudomonas Species: A Novel Study from Eutrophic Waters of Dal Lake, Kashmir, India. *Arch. Microbiol.* **2024**, *206*, 89. [CrossRef] [PubMed]
53. Shah, R.A.; Achyuthan, H.; Krishnan, H.; Lone, A.M.; Saju, S.; Ali, A.; Lone, S.A.; Malik, M.S.; Dash, C. Heavy Metal Concentration and Ecological Risk Assessment in Surface Sediments of Dal Lake, Kashmir Valley, Western Himalaya. *Arab. J. Geosci.* **2021**, *14*, 187. [CrossRef]
54. Paller, M.H.; Knox, A.S. Metal Bioavailability in Sediments and Its Role in Risk Assessment. In *Trace Elements in Waterlogged Soils and Sediments*; CRC Press: Boca Raton, FL, USA, 2024; pp. 267–285.
55. Perfus-Barbeoch, L.; Leonhardt, N.; Vavasseur, A.; Forestier, C. Heavy Metal Toxicity: Cadmium Permeates through Calcium Channels and Disturbs the Plant Water Status. *Plant J.* **2002**, *32*, 539–548. [CrossRef]

56. Mir, A.R.; Pichtel, J.; Hayat, S. Copper: Uptake, Toxicity and Tolerance in Plants and Management of Cu-Contaminated Soil. *BioMetals* **2021**, *34*, 737–759. [CrossRef]
57. Zulfiqar, U.; Haider, F.U.; Ahmad, M.; Hussain, S.; Maqsood, M.F.; Ishfaq, M.; Shahzad, B.; Waqas, M.M.; Ali, B.; Tayyab, M.N.; et al. Chromium Toxicity, Speciation, and Remediation Strategies in Soil-Plant Interface: A Critical Review. *Front. Plant Sci.* **2023**, *13*, 1081624. [CrossRef]
58. Hu, X.; Wei, X.; Ling, J.; Chen, J. Cobalt: An Essential Micronutrient for Plant Growth? *Front. Plant Sci.* **2021**, *12*, 768523. [CrossRef]
59. Pushnik, J.C.; Miller, G.W.; Manwaring, J.H. The Role of Iron in Higher Plant Chlorophyll Biosynthesis, Maintenance and Chloroplast Biogenesis. *J. Plant Nutr.* **1984**, *7*, 733–758. [CrossRef]
60. Kaur, H.; Kaur, H.; Kaur, H.; Srivastava, S. The Beneficial Roles of Trace and Ultratrace Elements in Plants. *Plant Growth Regul.* **2023**, *100*, 219–236. [CrossRef]
61. Abd Rasid, N.S.; Naim, M.N.; Che Man, H.; Abu Bakar, N.F.; Mokhtar, M.N. Evaluation of Surface Water Treated with Lotus Plant; Nelumbo Nucifera. *J. Environ. Chem. Eng.* **2019**, *7*, 103048. [CrossRef]
62. Liu, A.; Tian, D.; Xiang, Y.; Mo, H. Effects of Biochar on Growth of Asian Lotus (*Nelumbo nucifera* Gaertn.) and Cadmium Uptake in Artificially Cadmium-Polluted Water. *Sci. Hortic.* **2016**, *198*, 311–317. [CrossRef]
63. Painuly, A.S.; Gupta, R.; Vats, S. Bio-Accumulation of Arsenic (III) Using *Nelumbo nucifera* Gaertn. *J. Health Pollut.* **2019**, *9*, 190902. [CrossRef]
64. Wickramaratne, M.N.; Maduranga, T.M.; Chamara, L.S. Contamination of heavy metals in aquatic vegetables collected from cultivation sites in Sri Lanka. *J. Environ. Sci. Toxicol. Food Technol.* **2016**, *10*, 76–82.
65. Galal, T.M.; Farahat, E.A. The Invasive Macrophyte *Pistia stratiotes* L. as a Bioindicator for Water Pollution in Lake Mariut, Egypt. *Environ. Monit. Assess.* **2015**, *187*, 701. [CrossRef]
66. Bai, L.; Liu, X.L.; Hu, J.; Li, J.; Wang, Z.L.; Han, G.; Li, S.L.; Liu, C.Q. Heavy Metal Accumulation in Common Aquatic Plants in Rivers and Lakes in the Taihu Basin. *Int. J. Environ. Res. Public. Health* **2018**, *15*, 2857. [CrossRef]
67. Skorbiłowicz, E.; Skorbiłowicz, M.; Sidoruk, M. The Bioaccumulation of Potentially Toxic Elements in the Organs of *Phragmites australis* and Their Application as Indicators of Pollution (Bug River, Poland). *Water* **2024**, *16*, 3294. [CrossRef]
68. Yao, H.; Guo, L.; Jiang, B.H.; Luo, J.; Shi, X. Oxidative Stress and Chromium(VI) Carcinogenesis. *J. Environ. Pathol. Toxicol. Oncol.* **2008**, *27*, 77–88. [CrossRef] [PubMed]
69. Wang, S.; Zheng, Q.; Liu, Y.; Ze, Y.; Wang, Y.; Wu, Y.; Qi, J.; Qu, J.; Zhang, Y. Co-Incorporation of Nitrogen and Boron into Microscale Zero-Valent Iron via Mechanochemical Ball-Milling Method Improved Cr(VI) Elimination: Performance and Mechanism Investigation. *Chem. Eng. J.* **2025**, *506*, 160050. [CrossRef]
70. Megremi, I. Distribution and Bioavailability of Cr in Central Euboea, Greece. *Cent. Eur. J. Geosci.* **2010**, *2*, 103–123. [CrossRef]
71. Johri, N.; Jacquillet, G.; Unwin, R. Heavy Metal Poisoning: The Effects of Cadmium on the Kidney. *BioMetals* **2010**, *23*, 783–792. [CrossRef]
72. Gaetke, L.M.; Chow, C.K. Copper Toxicity, Oxidative Stress, and Antioxidant Nutrients. *Toxicology* **2003**, *189*, 147–163. [CrossRef] [PubMed]
73. Puntarulo, S. Iron, Oxidative Stress and Human Health. *Mol. Asp. Med.* **2005**, *26*, 299–312. [CrossRef]
74. Horning, K.J.; Caito, S.W.; Tipps, K.G.; Bowman, A.B.; Aschner, M. Manganese Is Essential for Neuronal Health. *Annu. Rev. Nutr.* **2015**, *35*, 71–108. [CrossRef]
75. Lippmann, M.; Ito, K.; Hwang, J.S.; Maciejczyk, P.; Chen, L.C. Cardiovascular Effects of Nickel in Ambient Air. *Environ. Health Perspect.* **2006**, *114*, 1662–1669. [CrossRef] [PubMed]
76. Dolomatov, S.I.; Sataeva, T.P.; Zukow, W. Modern Aspects of Regulatory, Pathophysiological and Toxic Effects of Cobalt Ions during Oral Intake in the Human Body. *Health Risk Anal.* **2019**, *2019*, 161–174. [CrossRef]
77. Osredkar, J. Copper and Zinc, Biological Role and Significance of Copper/Zinc Imbalance. *J. Clin. Toxicol.* **2011**, *s3*, 495. [CrossRef]
78. WHO/USEPA. Tolerable Daily Intake (TDI) Values Developed by International Organizations; 2016. Available online: https://semspub.epa.gov/work/HQ/190106.pdf (accessed on 10 January 2025).
79. Ologundudu, F.; Tobi, A.; Fopeyemi, O. Bioaccumulation Potential and Health Risk Assessment of Heavy Metals in *Corchorus olitorius* L. (Malvaceae) and *Amaranthus hybridus* L. (Amaranthaceae) Obtained from a Selected Dump Site in Akure, Nigeria. *Braz. J. Biol. Sci.* **2019**, *6*, 149–160. [CrossRef]
80. Obasi, P.N.; Chibuike, A.; Immaculate, N. Contamination Levels and Health Risk Assessment of Heavy Metals in Food Crops in Ishiagu Area, Lower Benue Trough South-Eastern Nigeria. *Int. J. Environ. Sci. Technol.* **2023**, *20*, 12069–12088. [CrossRef]

toxics

MDPI

Article

Modeling Desorption Rates and Background Concentrations of Heavy Metals Using a One-Dimensional Approach

Wendy Tatiana Gonzalez Cano [1,2,3], Serguei Lonin [3] and Kyoungrean Kim [1,2,*]

1 Marine Environmental Research Department, Korea Institute of Ocean Science and Technology, Busan 49111, Republic of Korea; tatiana@kiost.ac.kr
2 KIOST School, University of Science and Technology (UST), Busan 49111, Republic of Korea
3 Admiral Padilla Naval Academy of Cadets, Isla Naval Manzanillo, Cartagena de Indias 130001, Colombia; oceanmet.ltda@yahoo.com
* Correspondence: kyoungrean@kiost.ac.kr; Tel.: +82-51-664-3198

Abstract: Harmful heavy metals (HHMs) in marine sediments pose significant ecological and human health risks. This research developed a novel one-dimensional mathematical model to investigate the desorption rates and background concentrations (C_{bg}) of HHMs in cohesive sediments of coastal environments, using Cartagena Bay (CB), Colombia, as a reference for estuarine systems. The model integrates mass balance and molecular diffusion equations incorporating porosity and tortuosity. Both the particulate and dissolved phases of HHMs were considered. Numerical experiments were conducted over 28 years with a daily time step, simulating four primary hydrodynamic processes: molecular diffusion, desorption, sedimentation, and turbulent water exchange. The spatiotemporal evolution of C_{bg} provides valuable insights for sediment modeling, policy development, and advancing the understanding of HHM pollution in sediments. Results of the model align closely with empirical data from CB, demonstrating its applicability not only to local conditions but also to similar contaminated areas through a generalized approach. This model can be used as a reliable computational tool for managing coastal environments.

Keywords: estuarine sediments; heavy metals; background concentrations; desorption rate; cohesive sediment transport; mathematical modeling

1. Introduction

Heavy metal pollution poses a global environmental concern due to its severe toxicity [1,2], long-term persistence, and bioaccumulation in food chains [3–5]. Harmful heavy metals (HHMs) are continuously introduced into the environment [6] through natural and anthropogenic sources [7,8]. In estuarine waters, the presence of HHMs is generally observed in two distinct phases: dissolved in the water column and particulate adsorbed on the sediments. The partitioning of HHMs between these phases depends on the physical and chemical characteristics of suspended particles [9,10], in conjunction with environmental conditions such as salinity, pH, and dissolved organic matter [10].

Upon entering surface waters, HHMs are transported by rivers through wash load transport and eventually accumulate in marine sediments. Hydro-sedimentary processes such as desorption [11], resuspension, and dredging can release these contaminants back into the overlying water column [12], affecting water quality, the marine environment, and human health. Although HHM adsorption initially occurs near areas with significant anthropogenic activities, this study emphasizes the downstream consequences, particularly focusing on sediment contamination.

 https://doi.org/10.3390/toxics13060421

Various researchers rely on predicting interactions between water and sediments as a critical method for understanding HHM pollution [5,13–15]. Mathematical models have become powerful tools to address complex research questions related to coastal environments, offering reliable, cost-effective, and time-saving approaches [16].

Specifically, reaction-transport models, as described by Boudreau [17], Lynch and Officer [18], and Nicolis [19], have been pivotal in advancing sediment diagenesis and biogeochemical modeling, integrating key processes such as molecular diffusion, advection, and chemical reactions. These frameworks form the theoretical foundation of this study, focused on the vertical distribution and temporal evolution of HHMs in sediments.

Previous studies, such as Wu et al. [10], developed a two-dimensional (2D) transport model, later integrated into a one-dimensional (1D) framework to simulate the movement of dissolved and particulate HHMs along estuaries. However, these models did not explicitly address the accumulation of HHM contaminants or their subsequent phase evolution within the substrate, a significant gap in understanding the long-term impacts and interactions of HHMs with sediment dynamics.

Numerous studies have focused on HHM accumulation in sediments near contamination sources. However, limited studies have been published on desorption rates and their dynamics in downstream depositional environments such as lakes, lagoons, and estuaries. In systems dominated by wash load transport of fine cohesive sediments, the deposition and accumulation pathways of metal-bearing particles are not well documented. Apparently, sediment accumulation may occur slowly in distal areas, comparable to sorption rates, or rapidly near river mouths due to abrupt precipitation of materials, limiting metal exchange with the water column. These spatial heterogeneities affect HHM redistribution within sediments, complicating the estimation of background concentrations (C_{bg}).

This study advances existing frameworks by proposing a novel 1D model that couples transport and reaction processes. This model is designed for application at each computational node of a generalized three-dimensional (3D) hydrodynamic model, serving as a boundary condition at the water–sediment interface to simulate HHM accumulation and sediment evolution over time. The innovation of this approach is the integration of both dissolved and particulate HHM phases. Four critical hydrodynamic processes are quantified and modeled to evaluate their influence on HHM dynamics. This model, in accordance with empirical data, assumes that the dissolved-phase concentration (C_d) is considerably lower than the particulate-phase concentration (C_p) in the water column ($C_d << C_p$).

To demonstrate the model's applicability, simulations were referenced against estuarine conditions in the Cartagena Bay, Colombia (CB), a system subject to intense sedimentation and HHM inputs. Results obtained through this model align closely with empirical observations, reinforcing its validity.

The aim of this study is (1) to develop a novel 1D mathematical model to investigate HHM dynamics in estuarine sediments and (2) to elucidate the processes of HHMs governing background concentrations (C_{bg}). These concentrations serve as critical indicators for identifying anthropogenic inputs [20] and facilitate the formulation of effective management and remediation strategies [7,21]. This research represents the first attempt to establish the C_{bg} of HHMs in Colombia, highlighting its significance in addressing local and regional environmental concerns.

The model is applicable beyond the specific conditions of the Colombian coastline and could be effectively extended to various aquatic systems, including rivers, estuaries, and lakes worldwide affected by sediment contamination. Model outputs, including dynamic profiles illustrating the temporal drift of C_{bg}, are presented and critically discussed.

2. Materials and Methods

2.1. Cartagena Bay: A Reference System for Estuarine Conditions

CB is a semi-enclosed estuarine system on Colombia's Caribbean coast (Figure 1) (10°16′–10°26′ N, 75°30′–75°35′ W), with an average depth of 16 m, a maximum depth of 32 m, and a surface area of 84 km^2 [22]. The bay receives large amounts of sediments [23], nutrients, wastewater runoff [24], and contaminants from the Dique Channel [25,26], an artificial structure connected to the extensive Magdalena River basin (260,000-km^2) [26,27].

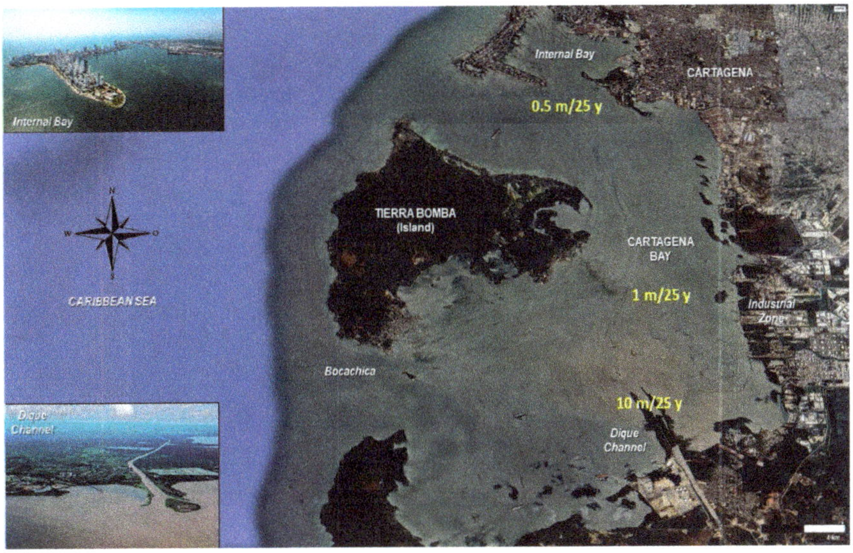

Figure 1. Location of the Cartagena Bay, Colombia (10°24′ N, 75°30′ W), showing the variability of sedimentation across the bay with three estimated deposition rates: Dique Channel mouth: 10 m/25 y; the central bay: 1 m/25 y; and the northern sector: 0.5 m/25 y. The Dique Channel is highlighted as the main sediment input pathway. Source: authors, modified from Google Earth; pictures taken from Google Images (accessed May 2025).

The Magdalena River transports a sediment flux of 184 Mt yr^{-1} and delivers the highest freshwater discharge (6496 m^3 s^{-1}) and sediment load (144 Mt yr^{-1}) to the Caribbean [26,27]. Seasonal rainfall from the Magdalena River, where the Dique Channel diverges, strongly influences the hydrology and sediment quality of CB [28]. CB's sedimentation patterns and morpho-dynamic characteristics have been previously studied through observations and modeling [29–31]. Due to wash load transport, HHM adsorption does not occur in CB but rather in distant sources before HHMs are transported downstream. This explains the HHM accumulation in precipitated sediments.

2.2. Mathematical Model

2.2.1. Definition of the Physical Problem

This research considers the wash load transport [32] of fine cohesive sediments (silt and clay) loaded with HHMs in particulate form. The primary source of HHMs is the rapid upstream industrialization in the Magdalena River. Particles are deposited at the bottom in the form of flocs. In the porous medium of precipitated sediments, desorption continues but at a lower rate than that during transport in the water column.

In the water column, HHMs exist in colloidal, particulate, and dissolved phases. The concentration of HHMs in dissolved form (C_d) is generally lower than that in particulate form (C_p). For instance, field measurements in the Magdalena River show that C_d values are approximately 1000 times lower than C_p in suspended and bottom sediments. This aligns with observations by Bartlett and Craig (1981) [33], who reported strong correlations (r = 0.94) between mercury (Hg) and silt in British estuaries, demonstrating Hg's affinity for fine particles and sulfur-rich organic matter (<0.0625 mm). The volumetric concentration of suspended sediments at the transition between dilute and concentrated systems is typically below 10^{-4} (dimensionless). Considering that desorption occurs over several years (~6 years), the mass of HHMs released from particulates is dispersed into a much larger water volume, further supporting the assumption $C_d << C_p$. Combined with the infinite-dilution diffusion concept [34], these theoretical and empirical insights justify the assumption as a valid simplification within this modeling framework.

Hereafter, we assume that C_d concentrations are multiplied by the constant K_d, which represents the equilibrium distribution coefficient. In the sediment substrate, these concentrations are also assumed to reach equilibrium due to limitations in molecular diffusion, which is partially restricted by porosity (n) and tortuosity (θ). Lower porosity and higher tortuosity restrict molecular diffusion, reducing HHM exchange with overlying waters and consequently promoting high accumulation and persistence within the sediment layer.

Tortuosity quantifies the complexity of pore pathways through which water and dissolved substances, such as HHMs, move within sediment layers [35]. Porosity, defined as the ratio of pore volume to total sediment volume [17], also plays a critical role in transport dynamics. Lower porosity implies fewer and smaller pores, restricting mobility and facilitating contaminant accumulation. As sediment compacts over time, porosity typically decreases with depth (z), becoming a time-dependent function. This leads to a gradual increase in substrate thickness in the absence of resuspension.

The desorption rate (γ), reflects the release of HHMs from C_p to C_d and depends on the grain diameter of sediments (d_{50}), their porosity (n), salinity (S), and pH. The porosity and tortuosity together influence molecular diffusion, calculated in the model using the Schmidt number (Sc), a dimensionless parameter used to characterize the relationship between the molecular viscosity of water and the diffusion of substances [36].

A 1D vertical model was formulated, neglecting the horizontal dispersion of HHMs, with the vertical axis directed upward from $z = 0$ (the reference level is assumed to be the starting point of sedimentation, as shown in Figure 2). The 1D model can be considered a sufficient approximation, considering that (a) the relationship between the vertical scale of the sediment layer and its horizontal extent along an estuary or river is small, and (b) exchange processes in the substrate in the vertical direction are much faster than the horizontal dynamics.

The domain is defined as $\{0 \le z \le D(t); t \ge 0\}$, where D = sediment thickness as a function of time (t). At the initial time, D was set as $D(t = 0) = 0$. To avoid singularities when solving the differential equations of the model, we assumed that $\frac{\partial D}{\partial t} > 0, \forall t$ under the absence of resuspension. The dynamics of the layer $D(t)$ are expressed as follows:

$$\frac{\partial D}{\partial t} = -w_g C_v (1-n)^{-1} = \frac{-w_g (1-n)^{-1} C_m}{\rho_S}, \quad (1)$$

where w_g is the settling velocity of sediments due to gravity, given by the Stokes formula ($w_g < 0$), and C_v, C_m, and ρ_s are the volumetric and mass concentrations of suspended sediments and their density, respectively. Within the bottom substrate, the molecular diffusion flux of the C_d is defined as follows:

$$Q = \alpha_S \nu \frac{\partial C_d}{\partial z}, \quad (2)$$

where ν is the kinematic molecular viscosity of water (Constant) and α_S defines the inverse Schmidt number ($\alpha_S = Sc^{-1}$) [36], which generally depends on time and substrate level or porosity n.

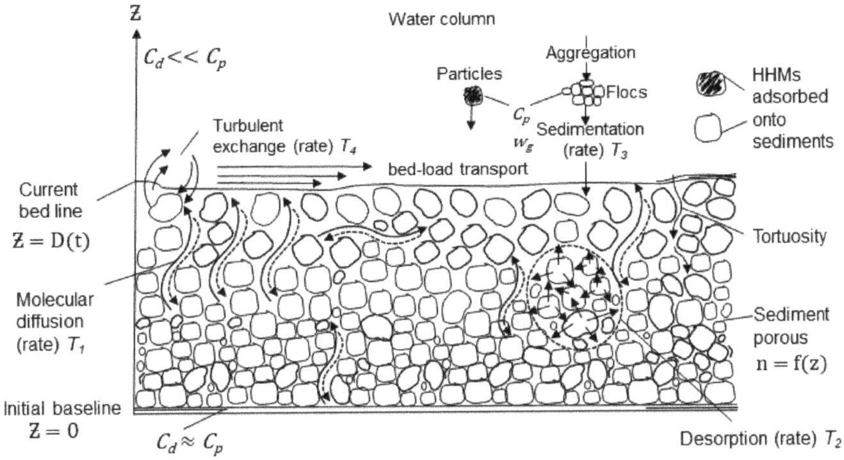

Figure 2. Conceptual model of HHM dynamics at the water–sediment interface and within the sediment substrate.

To determine the desorption rate $\gamma = f(d_{50}, n, S, pH)$, at least three timescales must be considered: (1) the molecular diffusion rate (T_1) of HHMs; (2) the desorption rate (T_2); and (3) the sediment deposition rate (T_3) at the bottom, as follows:

$$T_1 = \frac{D_0^2 Sc}{\nu}, \tag{3}$$

$$T_2 = \frac{1}{2\gamma}, \tag{4}$$

$$T_3 = \frac{D_0}{w_g C_v}. \tag{5}$$

This study has focused on the formulation and evaluation of a generalized 1D model for simulating HHM behavior in estuarine sediments, rather than on site-specific applications. The parameters of the model, particularly sedimentation and desorption rates, were calibrated within observed HHM concentrations reported in empirical data from CB [24,25,27,37]. The mathematical model was implemented and numerically solved using FORTRAN 90. Model outputs were compared to published sediment data from CB [23–25,38]. Simulated C_{bg} ranging from 1.0–2.4 mg kg^{-1} dw closely matched these empirical values at the sediment base (z = 0). The median grain diameter (d_{50}) (Table 1) was measured using a laser diffraction particle size analyzer. HHM concentrations were determined in the collected data using standard laboratory procedures involving acid digestion followed by quantification via atomic absorption spectrometry (AAS) or inductively coupled plasma mass spectrometry (ICP-MS). Although full calibration was limited by data availability, alignment with observed depth-integrated values provides partial validation.

Table 1. State variables and parameters used in this study.

Parameter	Description	Unit	Value	Reference
C_{bg}	HHM background concentration	g L^{-1}, mg kg^{-1} (dw) *	see Table 2	calculated
C_D	Drag coefficient	/	2×10^{-3}	[39]
C_d	Dissolved-phase HHM concentration	g L^{-1}, mg kg^{-1} (dw) *	/	calculated
C_m	Suspended-sediment mass concentration	g L^{-1}	/	[40]
C_p	Particulate-phase HHM concentration	g L^{-1}, mg kg^{-1} (dw) *	/	calculated
C_{p0}	Initial particulate HHM at precipitation	g L^{-1}	/	assumed
C_v	Suspended-sediment volumetric concentration	/	10^{-4}–10^{-5}	assumed
d_{50}	Median grain diameter of sediment	m	/	measured
D	Sediment thickness	m	0–1.6 **	calculated
F	Porosity–tortuosity factor	/	/	calculated
$HHMs$	Harmful heavy metals	g L^{-1}, mg kg^{-1} (dw) *	varies	measured
K_d	Coefficient of equilibrium distribution	/	/	assumed
m	Exponent in the relationship of Sc and n	/	/	literature
N	Number of computational nodes	/	100	assumed
n	Porosity	/	0.4	[34]
Q	Molecular diffusion flux	kg m^{-2} s^{-1}	varies	calculated
S	Salinity	/	0.06–35.7	assumed
S_C	Schmidt number	/	10–100	[6]
t	Time	s	0–8.64×10^8 s	assumed
T_1	Molecular diffusion rate	yr	0.3–3	calculated
T_2	Desorption rate	yr	3.15 (for $\gamma = 5 \times 10^{-8}$)	calculated
T_3	Sediment rate	yr	>31	calculated
T_4	Turbulent exchange rate	yr	/	calculated
u_*	Friction (dynamic) velocity	m s^{-1}	0–0.01	assumed
w_g	Settling velocity of sediments due to gravity	m s^{-1}	10^{-5}	assumed
Y	Dimensionless vertical coordinate	/	0–1	calculated
z	Vertical level within the substrate	m	0–1.6	calculated
z_0	Roughness parameter	m	/	literature
α_S	Inverse Schmidt number $\left(Sc^{-1}\right)$	/	0.01–0.1	[36]
γ	Desorption coefficient	s^{-1}	5×10^{-8}–1×10^{-9}	[41]
Δy	Vertical grid size in dimensionless coordinates	/	$1/(N-1)$	calculated
θ	Tortuosity	/	/	[35]
κ	Karman constant	/	0.41	literature
ν	Kinematic molecular viscosity of water	m^2 s^{-1}	10^{-6}	constant
ρ_s	Sediment–particle density	kg m^{-3}	2650	[39]
χ_0	Molecular diffusion coefficient (water only)	m^2 s^{-1}	/	[17]
χ_S	Molecular diffusion coefficients (with sediments)	m^2 s^{-1}	/	calculated

* C_d and C_p represent concentrations expressed in g L^{-1} or kg m^{-3} in the model for consistency, but in the figures, they are presented in mg kg^{-1} dry weight (dw) for easier comparison with laboratory data. Laboratories generally measure HHM concentrations in mg kg^{-1}, dw. Note that this difference in units is important when interpreting model results and comparing them with laboratory data or figures. ** D (0–1.6 m) based on 28 years of sedimentation. "/" means no value.

Four numerical simulation scenarios (Cases 1–4) were analyzed to investigate the influence of hydrodynamic parameters on HHM accumulation dynamics. Case 1 ($\gamma = 5 \times 10^{-8}$ s^{-1}) represents relatively fast desorption conditions, whereas Case 2 ($\gamma = 10^{-8}$ s^{-1}) examines the system response under slower desorption dynamics. Case 3 simulates a time-dependent increase in the C_p of HHMs, linearly increasing from 0 to 2.4 mg kg^{-1} over 28 years, reflecting observed historical contamination trends from distant sources such as the Magdalena River. Case 4 incorporates variable sediment inputs (55–250 m^3 s^{-1}), modeled through stochastic annual fluctuations to replicate seasonal variations typical of the Dique Channel. These scenarios evaluate sedimentation and desorption processes under contrasting environmental conditions with broader applicability.

Table 2. Model-derived Hg background concentrations (C_{bg}) at boundary (z = 0) under different simulation cases.

Case	Description	Estimated C_{bg} mg kg^{-1} (dw)	Observation
1	$\gamma = 5 \times 10^{-8}$ 1/s	1.4–1.7	Long-term equilibrium at $z = 0$
2	$\gamma = 10^{-8}$ 1/s	1.0–1.2	Slower equilibrium from low γ
3	C_{p0} increasing over 28 yr	2.0–2.4	Closest to observed CB field data
4	Variable sediment input *	2.0–2.2	Dynamic but consistent C_{bg} at $z = 0$
-	Average Hg C_{bg} (model)	0.2 ± 1.7	Variability across all cases

* Seasonal sediment variation using white noise flow 55–250 m^3 s^{-1}.

2.2.2. Governing Equations and Boundary Conditions

The mathematical formulation of the problem is expressed in Equation (1). The governing equations for C_p and C_d of HHMs are defined clearly below (Equations (6) and (7)), including mass balance constraints and desorption processes:

$$\frac{\partial C_p}{\partial t} = -\gamma(C_p - C_d) - \frac{w_g}{D} C_{p0}\delta(z - D) \tag{6}$$

$$\frac{\partial C_d}{\partial t} = \gamma(C_p - C_d) + \frac{\partial}{\partial z}\left(\alpha_S \nu \frac{\partial C_d}{\partial z}\right), \tag{7}$$

$$AC_d + \alpha_S \nu \frac{\partial C_d}{\partial z} = 0, \text{ at } z = D(t) \tag{8}$$

$$\frac{\partial C_d}{\partial z} = 0, \text{ at } z = 0 \tag{9}$$

In Equation (8), A is a constant defined in Appendix B, based on the fact that in the water column, $C_d \ll C_p$ due to diffusion in open water systems. As stated by the no-flux condition in Equation (9), an asymptotic equilibrium is assumed at $z = 0$ between C_p and C_d, with values equal to the C_{bg} to be defined in this study.

This boundary condition assumes equilibrium at lower sediment layers ($z = 0$), reaching a balance due to decreased porosity and restricted molecular diffusion over time. In Equation (9), the molecular flux of C_d at $z = 0$ is assumed to be zero. The boundary conditions at $z = 0$ and $z = D(t)$ ensure the mass balance of HHMs within the sediment, accurately representing fluxes and the conservation of mass. At $z = D(t)$, the boundary condition models the exchange between the sediment and the overlying water column.

2.2.3. Numerical Solution Under Variable Boundary Conditions

Equations (6)–(9), with their respective initial conditions of $C_p(z,\ t = 0) = C_d(z, t = 0) = 0$, have a variable boundary at an initial thickness of $D(t = 0) = 0$. The vertical coordinate (z) was transformed into a non-dimensional coordinate, following Yao et al. [13], to improve numerical solution robustness. The system herein was reformulated using a new variable:

$$y = \frac{z}{D(t)}$$

This becomes $y_j = (j-1)\Delta y;\ j = 1, \ldots, N;\ \Delta y = \frac{1}{N-1}$, where N is the number of vertical computational nodes. Thus, combining Equations (6) and (7) with Equation (1), we obtain the following:

$$\frac{\partial C_p}{\partial t} + \frac{y}{D}\frac{w_g C_V}{(1-n)}\frac{\partial C_p}{\partial y} = -\gamma(C_p - C_d) - \frac{w_g}{D} C_{p0}\delta(z - D), \tag{10}$$

$$\frac{\partial C_d}{\partial t} + \frac{y}{D}\frac{w_g C_V}{(1-n)}\frac{\partial C_d}{\partial y} = \gamma(C_p - C_d) + \frac{v}{D^2}\frac{\partial}{\partial y}\left(\alpha_S \frac{\partial C_d}{\partial y}\right). \quad (11)$$

These equations were then discretized using an implicit time scheme, ensuring numerical stability regardless of the sediment thickness, D. The first derivatives concerning y were then represented using an "upward" scheme of $O(\Delta y^1)$. The solution was obtained using the Thomas factorization algorithm.

3. Results

3.1. Estimation of Molecular Diffusion (T_1), Desorption (T_2), and Sedimentation (T_3)

To estimate the timescales (T_1, T_2, T_3) given by Equations (3)–(5), a characteristic sediment thickness (D_0 = 1 m), molecular viscosity ($v = 10^{-6}$ m^2 s^{-1}), and Schmidt number (Sc = 100) were adopted [6]. With these parameters, T_1 was estimated to be 3 years. For Sc = 10, T_1 was approximately 115 days. According to Liu et al. [42], the values of γ vary between 10^{-8} and 10^{-9} s^{-1}. For $\gamma = 5 \times 10^{-8}$ s^{-1}, the timescale of T_2 was 3.15 years.

Finally, assuming that the volumetric concentration was between 10^{-4} and 10^{-5} and the settling velocity due to gravity was 10^{-5} m/s, the value of T_3 was greater than 31 years. Therefore, $T_3 >> \max(T_1, T_2)$ T_3 has the slowest timescale, while the other two timescales were similar to each other.

In addition to previously defined timescales (T_1–T_3), a fourth timescale (T_4), representing the turbulent water column exchange of dissolved HHMs at the sediment–water interface, was determined. Resuspension was not considered, as particulate-bound HHMs do not significantly participate in this exchange.

3.2. Numerical Experiments

The sediment density (ρ_s) was set at 2650 kg m^{-3}, with porosity (n) fixed at 0.4. A drag coefficient (C_D) of 2×10^{-3} [39] and an initial dynamic velocity of 0.01 m s^{-1} were applied. Numerical experiments were conducted over 28 years, using a daily time step (1 d). Figure 3 illustrates the temporal evolution of C_d and C_p from the beginning of precipitation on both the surface and base when the desorption coefficient γ is varied. Sedimentation was assumed to continue uniformly over the 28-year simulation at constant rates, with fixed HHM concentrations in the precipitated sediments. In Figure 4, profiles of HHM concentrations in sediments are presented at both the midpoint and the end of the numerical experiments.

Over 28 years, the sediment layer grew to 1.6 m, which aligned well with data from CB and served as a reference for this study. Following an initial transient period (Figure 3), the C_d and C_p stabilized. The desorption rate was slower, corresponding to 6.3 years on the timescale of this process, compared to the reference value of 3.15 years.

The vertical profiles exhibited an exponential variation in the upper layer of the substrate (Figure 4), over 30–40 cm, followed by a uniform distribution. The variation was attributed to the vertical molecular diffusion of HHMs and their loss, particularly in C_d, due to turbulent exchange with the water column at the bottom. The uniform distribution in the lower sublayer indicates equilibrium between the two phases; however, this equilibrium was not constant (Figure 4). Equilibrium stability between C_d and C_p is crucial for ecological risk assessments, as it governs HHM bioavailability and potential toxicity in benthic ecosystems [43].

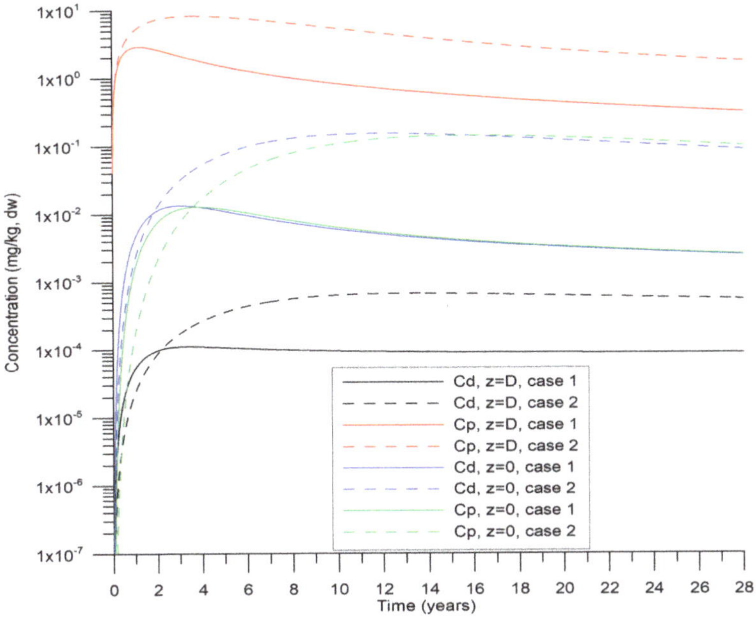

Figure 3. Temporal evolution of particulate (C_p) and dissolved (C_d) HHM concentrations at the variable bed level $D(t)$ and the basal level (z = 0) of the bottom substrate for Cases 1 and 2.

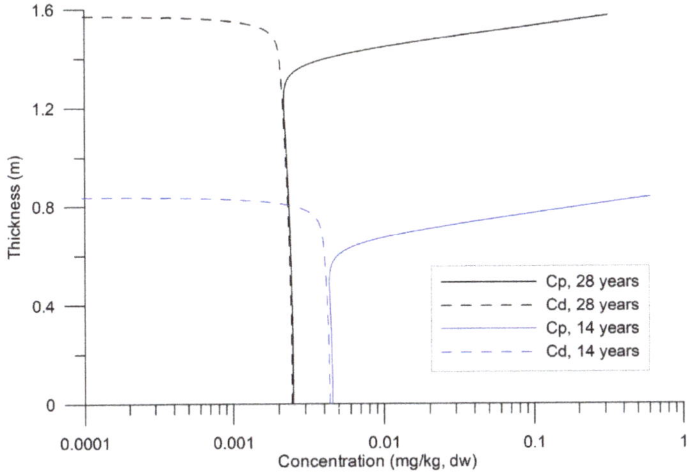

Figure 4. Temporal variability in the vertical profiles of particulate (C_p) and dissolved (C_d) HHMs for 14 and 28 years of sedimentation ($\gamma = 5 \times 10^{-8}\ s^{-1}$).

4. Discussion

4.1. Temporal Evolution of Background Concentrations Estimated by the Model

A drift value, tentatively called the background C_{bg}, was observed, characterized by a gradual decrease over time. This decrease is attributable to continuous slow molecular diffusion within the non-zero sediment porosity, transporting HHMs towards the sediment–water interface. Subsequent experiments were conducted by minimizing the turbulent exchange of the C_d with the water column above the bed. This occurs when $u_* \rightarrow 0$

(Figure 5). A nearly uniform distribution of C_d in the vertical direction was observed, along with the input of C_p at the bottom surface. The total concentrations of HHMs in the sediments were the sum of C_p and C_d/K_d, and in the laboratory, a single value was defined: "HHM concentrations in sediments".

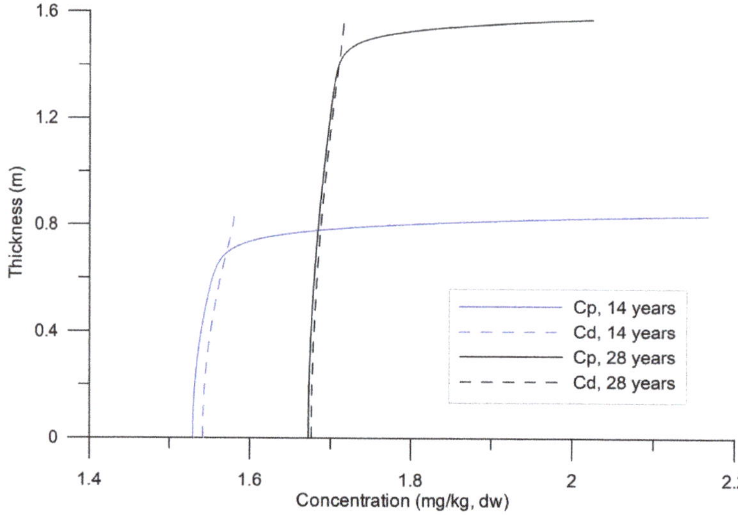

Figure 5. HHM concentrations in sediments under conditions of limited exchange between the dissolved phase and the water column.

For Case 1, γ was set to $5 \times 10^{-8}\ s^{-1}$, which suggested a relatively faster desorption rate compared to Case 2 ($\gamma = 10^{-8}\ s^{-1}$). Two additional cases (Cases 3 and 4) were simulated (Figure 6). In Case 3, the initial particulate concentration C_{p0}, increased linearly from 0 to 2.4 mg kg^{-1} over 28 years, reflecting observed trends in CB associated with increased HHM loading from a distant source, the Magdalena River. Case 4 was similar to Case 3, but with a variable sediment input that varied between 55 and 250 $m^3\ s^{-1}$ to replicate the Dique Channel's seasonal flow, using annually periodic white-noise perturbations (stochastic values 0–1).

The cases in Figure 6 were compared to Case 1 in Figure 3 where HHMs in sediments accumulated more slowly. Notably, when HHM loading gradually increased (Case 3), the concentrations at the sediment base ($z = 0$) consistently reached equilibrium ($C_d = C_p = C_{bg}$). Conversely, seasonal variations in sediment load from the Dique Channel (Case 4) did not significantly alter the equilibrium C_{bg} value. These findings imply that C_{bg} values remain stable despite short-term fluctuations, highlighting their value as robust indicators for long-term ecological risk assessments.

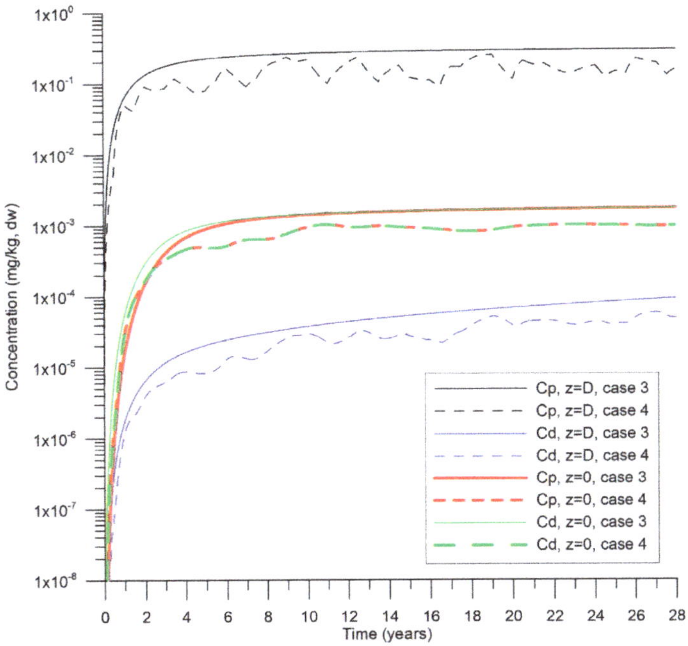

Figure 6. Temporal evolution of particulate (C_p) and dissolved (C_d) HHM concentrations at the variable bed level $D(t)$ and at the basal level (z = 0) of the bottom substrate for Cases 3 and 4.

4.2. Dimensionless Analysis and HHM Dynamics

To perform an analysis of the systems in Equations (1) and (6)–(9), dimensionless variables were introduced as follows:

$$\tilde{C}_d = \frac{C_d}{C_{bg}}; \tilde{C}_p = \frac{C_p}{C_{bg}}; \tilde{t}_1 = \frac{t}{T_1}; \tilde{t}_2 = \frac{t}{T_2}; \tilde{t}_3 = \frac{t}{T_3}; \tilde{z} = \frac{z}{D}; \frac{\tilde{t}_2}{\tilde{t}_1} = \frac{2\gamma D^2}{\alpha_S \nu} = a;$$
$$\tilde{C}_{p0} = \frac{C_{p0}}{C_{bg}}; \frac{\tilde{t}_1}{\tilde{t}_3} = \frac{w_g D}{\alpha_S \nu} = \beta; \tilde{D} = \frac{D}{D_0}; \zeta = \frac{\alpha_S \nu}{u_* D}; \tilde{C}_{p0} = \frac{C_{p0}}{C_{bg}}.$$

The "~" symbol implies a dimensionless variable.

The systems originally presented in Equations (6) and (7), along with their respective conditions (8) and (9), are reformulated in dimensionless form and have been renumbered consecutively as Equations (12)–(15), as follows:

$$\frac{\partial \tilde{C}_p}{\partial \tilde{t}_1} = -a\left(\tilde{C}_p - \tilde{C}_d\right) - \beta \tilde{C}_{p0} \delta\left(\tilde{z} - \tilde{D}\right), \tag{12}$$

$$\frac{\partial \tilde{C}_d}{\partial \tilde{t}_1} = a\left(\tilde{C}_p - \tilde{C}_d\right) + \frac{\partial}{\partial \tilde{z}}\left(\tilde{\alpha}_S \frac{\partial \tilde{C}_d}{\partial \tilde{z}}\right), \tag{13}$$

$$C_D^{\frac{1}{2}} \tilde{C}_d + \zeta \frac{\partial \tilde{C}_d}{\partial \tilde{z}} = 0,\ at\ \tilde{z} = \tilde{D} \tag{14}$$

$$\frac{\partial \tilde{C}_d}{\partial \tilde{z}} = 0,\ at\ \tilde{z} = 0. \tag{15}$$

Adding Systems (12) and (13) and using conditions (14) and (15), the results are as follows:

$$\int_{0}^{\tilde{D}} \frac{\partial\left(\tilde{C}_p + \tilde{C}_d\right)}{\partial \tilde{t}_1} d\tilde{z} = |\beta|\tilde{C}_{p0} - \frac{c_{\tilde{D}}^{\frac{1}{2}} \tilde{C}_d\left(\tilde{z} = \tilde{D}\right)}{\zeta}. \quad (16)$$

Applying Leibniz's rule and reformulating Equation (1) in terms of "fast" time $\tilde{t}_1$, we obtain the following:

$$\frac{\partial \tilde{D}}{\partial \tilde{t}_1} = \frac{C_v}{\{|\beta|(1-n)\}} \quad (17)$$

The temporal variation in the total HHM concentration of sediments over the entire extent of its layers can be defined by the following equation:

$$\frac{\partial}{\partial \tilde{t}_1} \int_{0}^{\tilde{D}} \left(\tilde{C}_p + \tilde{C}_d\right) d\tilde{z} = |\beta|\tilde{C}_{p0} - \frac{c_{\tilde{D}}^{\frac{1}{2}} \tilde{C}_d\left(\tilde{z} = \tilde{D}\right)}{\zeta} + \frac{C_v}{\{|\beta|(1-n)\}\left(\tilde{C}_p + \tilde{C}_d\right)\Big|_{\tilde{D}}}. \quad (18)$$

The first term on the right-hand side of Equation (18) represents the input of HHMs in C_p into the sediment column, while the second term accounts for their loss through exchange with the overlying water in the C_d. The final term corresponds to the increase in the total HHM concentration due to changes in substrate thickness and its redistribution in the column. This term was considered less relevant when the T_3 scale represented a slow time relative to T_1 and T_2. Within the same body of water, as exemplified by CB, the T_3 scale is spatially variable.

If $\frac{\partial \tilde{D}}{\partial \tilde{t}_1} \approx 0$ for the "fast" time in System (17), then systems (12)–(15) are represented as parabolic equations whose asymptotes in time are $\tilde{C}_p = \tilde{C}_d$. Steady-state conditions are possible only if the particle sedimentation process stops.

For $\rightarrow \infty$, $\frac{\partial}{\partial \tilde{z}}\left(\tilde{\alpha}_S \frac{\partial \tilde{C}_d}{\partial \tilde{z}}\right) = 0$, considering condition (15) at a given level, the molecular flow is equal to zero throughout the substrate.

In this case, $\tilde{C}_p = \tilde{C}_d = 1 \forall t$ (background concentration). The only reason this did not occur throughout the entire sediment column is the permanent entry of HHMs, owing to their precipitation on particles at the bottom and the exchange of the diluted phase with the water column at the same vertical level.

The simulated sedimentation rates ranged from 0.5 m per 25 years (low deposition) to 10 m per 25 years (at the river mouth). Figure 1 presents the variability in sedimentation across CB, highlighting three depositional zones: Dique Channel mouth: 10 m/25 y; the central bay: 1 m/25 y; and the northern sector: 0.5 m/25 y. Therefore, the 1D model should be applied at each calculation node of a 3D hydrodynamic mesh, with local scales adjusted accordingly.

The universality of the proposed model lies in its formulation using dimensionless variables and scale parameters. The analysis of dimensionless equations (Equations (12)–(15)) allows for a broad spectrum of environmental conditions. These ranges reflect both the intra-basin variability within CB, such as differences between river mouth and inner-bay sedimentation rates, and potential conditions in other estuarine systems. This dimensional analysis enables the model to be applied across geographically distinct water bodies, provided that the local sedimentation dynamics and hydro-sedimentary conditions are within comparable parameter bounds.

Considering the timescale variation in the main processes, the desorption rate, the average speed of sediment settling by gravity, and the molecular viscosity of water were fixed. The thickness of the sediment layer and its porosity (through the Schmidt number) varied within reasonable limits, characterizing CB as an example of an estuary. The resulting values of the dimensionless parameters for systems (12)–(15) were a = 10^{-3} to 10^{3}; $\beta = 10^{2}$ to 10^{4}; $\zeta/C_D^{\frac{1}{2}} = 0.25\ (10^{-1}–10^{-5})$; and $\tilde{\alpha}_S = 10^{-2}$ to 10^{2}.

Under these conditions, Equations (12) and (13) can present multiple scenarios of HHM dynamics because the ratios of scale $\frac{\tilde{t}_2}{\tilde{t}_1}$ and $\frac{\tilde{t}_1}{\tilde{t}_3}$ change four to six orders of magnitude. In the case where the parameter a = $\frac{\tilde{t}_2}{\tilde{t}_1}$, the scales become inverted. Regarding condition (14), the relationship $\zeta/C_D^{\frac{1}{2}} << 1$ implied an abrupt gradient $\frac{\partial \tilde{C}_d}{\partial \tilde{z}}$, which was observed in Figure 4 at the interface between sediments and water. This detail was not observed in the measurements of HHMs in sediments because the laboratories analyze the total concentrations, where $C_d/K_d + C_p$, and the C_p concentration predominates in the samples.

4.3. Model Assumptions, Limitations, and Ecological Implications

While CB served as a reference system to contextualize parameter ranges and model outputs, this study was not designed for site-specific application or empirical calibration. Rather, the model was developed to explore general physical processes governing HHM desorption and accumulation in estuarine sediments. Field data from CB, including reported sedimentation rates and Hg concentrations, were used qualitatively to guide parameter selection and verify that simulated C_{bg} remained within empirically observed values, supporting the model's realism under estuarine conditions.

The proposed 1D model was developed under the assumptions of continuous sedimentation without bottom erosion events. Technically, erosion could be easily included in the model; however, it may be difficult to control over extended periods of sediment dynamics. Changes in the porosity and tortuosity were also considered, which influence HHM transport and accumulation. These mechanisms may require a rheological model. Since the model operated under the assumption that the muddy substrate was not in motion, no assumption of the fluid type or the Newtonian fluid approximation is required.

A notable limitation of the current model is the assumption of a constant sedimentation velocity (w_g), whereas sedimentation processes in estuarine environments typically exhibit considerable spatial complexity. For instance, Figure 1 highlights sedimentation rates in CB varying by an order of magnitude between the Dique Channel mouth (10 m/25 y) and the northern bay sector (0.5 m/25 y). Such variations result from a combination of (a) bed load transport, (b) the precipitation of suspended particles, and (c) the flocculation of fine particles induced by brackish water salinity gradients. Determining the dominant mechanism among these and assessing the impact of wash load transport on sediment distribution throughout the bay remain challenging. Detailed geographic-specific analysis and further refinement of sedimentation mechanisms given in Equation (1) would thus enhance the model accuracy and applicability.

This modeling approach addresses a critical gap in the representation of sediment processes, particularly the understanding of C_{bg} of HHMs in estuarine sediments, by integrating transport and reaction processes with site-specific hydro-sedimentary influences, often simplified in traditional frameworks. The relevance of C_{bg} lies in its strong association with toxicological thresholds, bioavailability, and long-term ecological risks related to HHM pollution. Although a 1D framework offers notable advantages in computational efficiency, it limits horizontal transport and spatial interactions across estuarine gradients.

Future initiatives could benefit from integrating diagenetic and hydrodynamic models to support evidence-based environmental management for preserving estuarine ecosystems.

5. Conclusions

Under physically valid assumptions, a novel 1D mathematical model was developed to simulate HHM dynamics in estuarine sediments, with broad applicability to water bodies influenced by HHM contamination. This numerical framework advances prior approaches by integrating coupled transport-reaction processes while dynamically accounting for porosity and tortuosity. Unlike conventional models, this approach includes molecular diffusion (T_1), desorption (T_2), sedimentation (T_3), and water-turbulence exchange (T_4) as a distinct method to estimate HHM C_{bg}. A notable innovation is the separation of C_d and C_p, reaching an asymptotic equilibrium ($C_d = C_p = C_{bg}$) at the sediment base ($z = 0$). This mathematical formulation has not been previously reported in existing sediment models.

The C_{bg} for Hg in CB ranged from 1.0 to 2.4 mg kg^{-1} dw, providing a valuable reference for future ecological risk assessments, pollution indexing, and numerical model calibration in estuarine sediments. C_{bg} of Hg were characterized by very slow desorption. Particularly, C_{bg} values did not remain constant but exhibited a drift, influenced by limited exchange with upper layers and overlying water. These findings may improve ecological risk assessment, environmental monitoring, and policy formulation to mitigate HHM impacts in CB and similar contaminated ecosystems.

Spatial and temporal variability in C_{bg} arises from local sediment dynamics, precisely variations in precipitation rates, highlighting the need for zone-specific assessments within the same water body. Consequently, the 1D model can be applied to each node of the general hydrodynamic model of the basin.

The observed drift in C_{bg} values demonstrates that profiling sediment layers dated with 14C does not necessarily reflect historical in situ concentrations, as reported by Fukue et al. [44]. This issue draws attention and stimulates future research using inverse models to restore HHM ancient profiles from in situ measurements.

The 1D model would be implemented as an interface between the water column and the consolidated substrate. This intermediate layer would capture processes at the sediment–water interface, transitioning from Newtonian fluid properties in the water–sediment upper layer of the bottom to solid substrate characteristics. The model acts as a universal boundary condition applicable across diverse aquatic systems receiving HHM contamination. However, site-specific calibration may be necessary due to local sedimentological and hydrodynamic conditions.

Future works will focus on integrating the 1D model into a 3D hydrodynamic framework to continuously simulate the long-term fate of sediments and HHMs to compare the model's stratification predictions to in situ measurements of the vertical substrate. Such advancements could significantly aid in developing more effective management strategies to mitigate HHM pollution in coastal marine environments.

Author Contributions: Conceptualization, W.T.G.C., S.L. and K.K.; methodology, W.T.G.C., S.L. and K.K.; software, S.L.; validation, S.L. and K.K.; formal analysis, W.T.G.C.; investigation, W.T.G.C., S.L. and K.K.; resources, W.T.G.C., S.L. and K.K.; data curation, W.T.G.C.; writing—original draft preparation, W.T.G.C. and S.L.; writing—review and editing, S.L. and K.K.; visualization, W.T.G.C. and S.L.; supervision, K.K.; project administration, K.K.; funding acquisition, K.K. All authors have read and agreed to the published version of the manuscript.

Funding: This research was funded by the Korea Institute of Ocean Science and Technology, Republic of Korea, grant number PEA0301, and by the Ministry of Oceans and Fisheries, Republic of Korea, grant number KIMST-20220027. The authors gratefully acknowledge this support.

Institutional Review Board Statement: Not applicable.

Informed Consent Statement: Not applicable.

Data Availability Statement: Data available on request from the authors.

Acknowledgments: We thank everyone who was related to this research and the anonymous reviewers, whose valuable comments led to the enhancement of the final version of this manuscript. We would like to express our special gratitude to the Korea Institute of Ocean Science and Technology (KIOST), Republic of Korea, and the Colombian Navy for their support. Graphical abstract created in https://BioRender.com.

Conflicts of Interest: The authors declare no conflicts of interest.

Appendix A

Without the source term and molecular diffusion in a closed system, Equations (6) and (7) can be expressed as follows:

$$\frac{\partial C_p}{\partial t} = -\gamma(C_p - C_d) \tag{A1}$$

$$\frac{\partial C_d}{\partial t} = \gamma(C_p - C_d), \tag{A2}$$

This represents the desorption from the particulate to the dissolved form of the HHMs, with

$$\frac{\partial(C_p + C_d)}{\partial t} = 0$$

Assuming the initial conditions $C_p(t=0) = C_0$ and $C_d(t=0) = 0$, the respective analytical solution of the system (Equations (A1) and (A2)) becomes the following:

$$C_p(t) = \frac{C_0\left(1 + e^{-2\gamma t}\right)}{2};\, C_d(t) = \frac{C_0\left(1 - e^{-2\gamma t}\right)}{2}, \tag{A3}$$

until it reaches an equilibrium, where $C_p = C_d = \frac{C_0}{2}$ for $t \to \infty$. The characteristic scale of this process, given in Equation (4), according to Equation (A3) is $T_2 = \frac{1}{2\gamma}$.

Appendix B

To specify the boundary conditions for Equation (7) at the bottom surface, a flux equilibrium was established for the dissolved components of the HHMs:

$$-\overline{w'C'} + \nu\frac{\partial\overline{C}}{\partial z} = \alpha_S \nu \frac{\partial\overline{C_d}}{\partial z} \tag{A4}$$

Here, the first and second terms on the left-hand side represent the turbulent and molecular fluxes of the HHMs with concentrations $\overline{C}$ in the water, respectively. The term on the right-hand side is the molecular diffusion flux of the substance in the sediments.

The sum of the turbulent and molecular fluxes was constant in a relatively thick layer, known as the near-surface bulk layer or "layer of constant fluxes". This could be parameterized based on the K-theory of turbulence by defining the flux q within the logarithmic profile of the substance:

$$q = \frac{\left(\overline{C} - C_d\right)\kappa u_*}{ln\left(\frac{z}{z_0}\right)}, \tag{A5}$$

where κ = Karman constant ($\kappa = 0.41$); u_* is the friction (dynamic) velocity in the near-surface layer; z_0 is the roughness parameter. In this case, z extends from the bottom surface z_0 upwards and is fixed with a reference value $\overline{C}$ of concentrations.

Introducing the drag coefficient C_D and assuming that the concentration in water is $\overline{C} = 0$ in Equation (A5), and based $\overline{C} \ll C_d$, Equation (A4) gives the following:

$$-C_d C_D^{\frac{1}{2}} u_* = \alpha_S \nu \frac{\partial \overline{C_d}}{\partial z}, \quad \text{(A6)}$$

which is Equation (8) when $A = C_D^{1/2} u_*$.

The fact that $\overline{C} \ll C_d$ in the water column was justified under the assumption that the volumetric sediment concentrations in the water were less than 10^{-4}. In an extreme hypothetical scenario, where equilibrium is reached between the HHMs in the water column, encompassing both C_p and C_d, the concentration $\overline{C}$ represents no more than 0.01% of the C_p.

Appendix C

Generally, the molecular Schmidt number, Sc, characterizes the relationship between the molecular viscosity of water and the molecular diffusion of substances; it measures how fast the "diffusion of momentum" occurs relative to other fluid properties. Substances with a water temperature of approximately Sc = 10 can increase by one or two orders of magnitude [36].

The Schmidt number is associated with the porosity and tortuosity of a fluid composed of liquid and bottom sediments. Porosity and tortuosity are considered to play major roles in the increase of this number. In the proposed study by Maerki et al. [45], the molecular flow (2) was identified as follows:

$$Q = n\chi_S \frac{\partial C_d}{\partial z} = n\chi_0 \theta^{-2} \frac{\partial C_d}{\partial z} = F^{-1} \chi_0 \frac{\partial C_d}{\partial z}, \quad \text{(A7)}$$

where n represents porosity, θ characterizes tortuosity, and F combines the effect of both; χ_S and χ_0 are the molecular diffusion coefficients with and without sediment particles, respectively.

If $\chi_0 = \nu$ and all diffusion effects (substance, porosity, and tortuosity) are assigned to the Sc number, then $Sc = F$. Based on Maerki et al. [45], we conclude the following:

$$Sc = 1.02 n^{-m} \quad \text{(A8)}$$

where m > 1 (m = 1.81 in the cited work).

Consequently, the Sc value depends on the substrate level, residence time, and compaction of the sediment, among other factors. Shen and Chen [46] provided further details on the parameterizations of these effects in sediments.

References

1. Abdelmonem, B.H.; Kamal, L.T.; Elbaz, R.M.; Khalifa, M.R.; Abdelnaser, A. From Contamination to Detection: The Growing Threat of Heavy Metals. *Heliyon* **2025**, *11*, e41713. [CrossRef] [PubMed]
2. Proshad, R.; Asha, S.M.A.A.; Tan, R.; Lu, Y.; Abedin, M.A.; Ding, Z.; Zhang, S.; Li, Z.; Chen, G.; Zhao, Z. Machine Learning Models with Innovative Outlier Detection Techniques for Predicting Heavy Metal Contamination in Soils. *J. Hazard. Mater.* **2025**, *481*, 136536. [CrossRef] [PubMed]
3. Liu, H.; Ding, C.; Zhang, G.; Guo, Y.; Song, Y.; Thangaraj, S.; Zhang, X.; Sun, J. Dissolved and Particulate Heavy Metal Pollution Status in Seawater and Sedimentary Heavy Metals of the Bohai Bay. *Mar. Environ. Res.* **2023**, *191*, 106158. [CrossRef] [PubMed]
4. Liu, Q.; Yang, P.; Hu, Z.; Shu, Q.; Chen, Y. Identification of the Sources and Influencing Factors of the Spatial Variation of Heavy Metals in Surface Sediments along the Northern Jiangsu Coast. *Ecol. Indic.* **2022**, *137*, 108716. [CrossRef]

5. Zhou, L.; Wu, F.; Meng, Y.; Byrne, P.; Ghomshei, M.; Abbaspour, K.C. Modeling Transport and Fate of Heavy Metals at the Watershed Scale: State-of-the-Art and Future Directions. *Sci. Total Environ.* **2023**, *878*, 163087. [CrossRef]
6. Schnoor, J.; Sato, C.; McKechnie, D.; Sahoo, D. *Processes, Coefficients, and Models for Simulating Toxic Organics and Heavy Metals in Surface Waters*; EPA/600/3-87/015; Department of Civil and Environmental Engineering, University of Iowa: Iowa City, IA, USA, 1987.
7. Sun, Y.; Zhao, Y.; Hao, L.; Zhao, X.; Lu, J.; Shi, Y.; Ma, C.; Li, Q. Application of the Partial Least Square Regression Method in Determining the Natural Background of Soil Heavy Metals: A Case Study in the Songhua River Basin, China. *Sci. Total Environ.* **2024**, *918*, 170695. [CrossRef]
8. Luo, M.; Kang, X.; Liu, Q.; Yu, H.; Tao, Y.; Wang, H.; Niu, Y.; Niu, Y. Research on the Geochemical Background Values and Evolution Rules of Lake Sediments for Heavy Metals and Nutrients in the Eastern China Plain from 1937 to 2017. *J. Hazard. Mater.* **2022**, *436*, 129136. [CrossRef]
9. Kuang, Z.; Shi, Z.; Wang, H.; Du, S.; Gong, H.; Liu, Q.; Gu, Y.; Fan, Z.; Huang, H.; Wang, S. Bioavailability of Trace Metals in Sediments from Daya Bay Nature Reserve: Spatial Variation, Controlling Factors and the Exposure Risk Assessment for Aquatic Biota. *Ecol. Indic.* **2024**, *169*, 112789. [CrossRef]
10. Wu, Y.; Falconer, R.; Lin, B. Modelling Trace Metal Concentration Distributions in Estuarine Waters. *Estuar. Coast. Shelf Sci.* **2005**, *64*, 699–709. [CrossRef]
11. Chang, S.; Han, L.; Chen, R.; Liu, Z.; Fan, Y.; An, X.; Zhai, Y.; Wu, P.; Wang, T. Vulnerability Assessment of Soil Cadmium with Adsorption–Desorption Coupling Model. *Ecol. Indic.* **2023**, *146*, 109904. [CrossRef]
12. Ristea, E.; Pârvulescu, O.C.; Lavric, V.; Oros, A. Assessment of Heavy Metal Contamination of Seawater and Sediments Along the Romanian Black Sea Coast: Spatial Distribution and Environmental Implications. *Sustainability* **2025**, *17*, 2586. [CrossRef]
13. Yao, J.; Liu, W.; Chen, Z. Numerical Solution of a Moving Boundary Problem of One-Dimensional Flow in Semi-Infinite Long Porous Media with Threshold Pressure Gradient. *Math. Probl. Eng.* **2013**, *2013*, 384246. [CrossRef]
14. Bhandari, A.; Surampalli, R.Y.; Champagne, P.; Ong, S.K.; Tyagi, R.D.; Lo, I.M.C. *Remediation Technologies for Soils and Groundwater*; American Society of Civil Engineers (ASCE): Reston, VA, USA, 2007; Volume 60, pp. 1–449. [CrossRef]
15. Liang, H.Y.; Zhang, Y.H.; Du, S.L.; Cao, J.L.; Liu, Y.F.; Zhao, H.; Ding, T.T. Heavy Metals in Sediments of the River-Lake System in the Dianchi Basin, China: Their Pollution, Sources, and Risks. *Sci. Total Environ.* **2024**, *957*, 177652. [CrossRef]
16. Lonin, S.; Andrade, C.A.; Monroy, J. Wave Climate and the Effect of Induced Currents over the Barrier Reef of the Cays of Alburquerque Island, Colombia. *Sustainability* **2022**, *14*, 6069. [CrossRef]
17. Boudreau, B.P. *Diagenetic Models and Their Implementation*; Springer: Berlin/Heidelberg, Germany, 1998; Volume 15, ISBN 978-3-642-64399-6.
18. Lynch, D.R.; Officer, C.B. Nonlinear Parameter Estimation for Sediment Cores. *Chem. Geol.* **1984**, *44*, 203–225. [CrossRef]
19. Nicolis, C. Tracer Dynamics in Ocean Sediments and the Deciphering of Past Climates. *Math. Comput. Model.* **1995**, *21*, 27–38. [CrossRef]
20. Sun, Y.; Yang, J.; Li, K.; Gong, J.; Gao, J.; Wang, Z.; Cai, Y.; Zhao, K.; Hu, S.; Fu, Y.; et al. Differentiating Environmental Scenarios to Establish Geochemical Baseline Values for Heavy Metals in Soil: A Case Study of Hainan Island, China. *Sci. Total Environ.* **2023**, *898*, 165634. [CrossRef]
21. Jung, J.M.; Kim, C.J.; Chung, C.S.; Kim, T.; Gu, H.S.; Kim, H.E.; Choi, K.Y. Applying New Regional Background Concentration Criteria to Assess Heavy Metal Contamination in Deep-Sea Sediments at an Ocean Dumping Site, Republic of Korea. *Mar. Pollut. Bull.* **2024**, *200*, 116065. [CrossRef]
22. Olivero-Verbel, R.; Eljarrat, E.; Johnson-Restrepo, B. Organophosphate Ester Flame Retardants in Sediments and Marine Fish Species in Colombia: Occurrence, Distribution, and Implications for Human Risk Assessment. *Mar. Pollut. Bull.* **2025**, *213*, 117654. [CrossRef]
23. Caballero-Gallardo, K.; Alcala-Orozco, M.; Barraza-Quiroz, D.; De la Rosa, J.; Olivero-Verbel, J. Environmental Risks Associated with Trace Elements in Sediments from Cartagena Bay, an Industrialized Site at the Caribbean. *Chemosphere* **2020**, *242*, 125173. [CrossRef]
24. Tosic, M.; Restrepo, J.D.; Lonin, S.; Izquierdo, A.; Martins, F. Water and Sediment Quality in Cartagena Bay, Colombia: Seasonal Variability and Potential Impacts of Pollution. *Estuar. Coast. Shelf Sci.* **2019**, *216*, 187–203. [CrossRef]
25. Romero-Murillo, P.; Gallego, J.L.; Leignel, V. Marine Pollution and Advances in Biomonitoring in Cartagena Bay in the Colombian Caribbean. *Toxics* **2023**, *11*, 631. [CrossRef] [PubMed]
26. Tosic, M.; Ángel, J.D.R. Sustainability Impacts of Sediments on the Estuary, Ports, and Fishing Communities of Cartagena Bay, Colombian Caribbean. *Wiley Interdiscip. Rev. Water* **2023**, *11*, e1709. [CrossRef]
27. Cano, W.T.G.; Kim, K. How to Achieve Sustainably Beneficial Uses of Marine Sediments in Colombia ? *Sustainability* **2022**, *14*, 14821. [CrossRef]
28. Tosic, M.; Martins, F.; Lonin, S.; Izquierdo, A.; Restrepo, J.D. Hydrodynamic Modelling of a Polluted Tropical Bay: Assessment of Anthropogenic Impacts on Freshwater Runoff and Estuarine Water Renewal. *J. Environ. Manag.* **2019**, *236*, 695–714. [CrossRef]

29. Andrade, C.; Thomas, Y.F.; Lonin, S.; Parra, C.; Kunesch, S.; Menanteau, L.; Andriau, A.; Piñeres, C.; Velasco, S. Aspectos morfodinámicos de la bahía de Cartagena de Indias. *Bol. Cient. CIOH* **2004**, *22*, 90–104. [CrossRef]
30. Thomas, Y.F.; Cesaraccio, M.; Kunesch, S.; Andrieu, A.; Ménanteau, L.; Andrade, C.; Lonin, S.; Parra, C.; Pineres, C.; Velasco, S.P. Étude Morphodynamique de La Baie de Carthagène Des Indes, Colombie. In *Milieux Littoraux, Nouvelles Perspectives D' Études*; L'Harmattan: Paris, France, 2005; pp. 171–191.
31. Thomas, Y.F.; Ménanteau, L.; Kunesch, S.; Cesaraccio, M.; Andrade, C.; Lonin, S.; Parra, C. Le delta du canal du Dique (baie de Carthagène des Indes, Colombie). Modélisation géomorphologique et sédimentologique. In Proceedings of the Colloque International Interactions-Nature-Sociétés, Analyses et Modèles, La Baule-Escoublac, France, 3–6 May 2006; pp. 1–7.
32. Van Rijn, L.C. *Principles of Sediment Transport in Rivers, Estuaries and Coastal Seas*; Aqua Publications: Amsterdam, The Netherlands, 1993.
33. Bartlett, P.; Craig, P. Total Mercury and Methyl Mercury Levels in British Estuarine Sediments—II. *Water Res.* **1981**, *15*, 37–47. [CrossRef]
34. Boudreau, B.P. The Diffusive Tortuosity of Fine-Grained Unlithified Sediments. *Geochim. Cosmochim. Acta* **1996**, *60*, 3139–3142. [CrossRef]
35. Ghanbarian, B.; Hunt, A.G.; Ewing, R.P.; Sahimi, M. Tortuosity in Porous Media: A Critical Review. *Soil Sci. Soc. Am. J.* **2013**, *77*, 1461–1477. [CrossRef]
36. Monin, A.S.; Yaglom, A.M. *Statistical Fluid Mechanics: The Mechanics of Turbulence, Volume 1*; MIT Press: Cambridge, MA, USA, 1973; Volume 60.
37. Orani, A.M.; Vassileva, E.; Azemard, S.; Alonso-Hernandez, C. Trace Elements Contamination Assessment in Marine Sediments from Different Regions of the Caribbean Sea. *J. Hazard. Mater.* **2020**, *399*, 122934. [CrossRef] [PubMed]
38. Olivero-Verbel, J.; Caballero-Gallardo, K.; Torres-Fuentes, N. Assessment of Mercury in Muscle of Fish from Cartagena Bay, a Tropical Estuary at the North of Colombia. *Int. J. Environ. Health Res.* **2009**, *19*, 343–355. [CrossRef] [PubMed]
39. Marchuk, G.I.; Kagan, B.A. *Ocean Tides: Mathematical Models and Numerical Experiments*; Pergamon Press: Oxford, UK, 1984.
40. Pintilie, S.; Brânză, L.; Bețianu, C.; Pavel, L.V.; Ungureanu, F.; Gavrilescu, M. Modelling and Simulation of Heavy Metals Transport in Water and Sediments. *Environ. Eng. Manag. J.* **2007**, *6*, 153–161. [CrossRef]
41. Luo, P.; Luo, M.; Li, F.; Qi, X.; Huo, A.; Wang, Z.; He, B.; Takara, K.; Nover, D.; Wang, Y. Urban Flood Numerical Simulation: Research, Methods and Future Perspectives. *Environ. Model. Softw.* **2022**, *156*, 105478. [CrossRef]
42. Liu, Q.; Jia, Z.; Liu, G.; Li, S.; Hu, J. Assessment of Heavy Metals Remobilization and Release Risks at the Sediment-Water Interface in Estuarine Environment. *Mar. Pollut. Bull.* **2023**, *187*, 114517. [CrossRef]
43. He, L.; Chen, G.; Wang, X.; Shen, J.; Zhang, H.; Lin, Y.; Shen, Y.; Lang, F.; Gong, C. Pollution Characteristics and Risk Assessment of Heavy Metals in the Sediments of the Inflow Rivers of Dianchi Lake, China. *Toxics* **2024**, *12*, 322. [CrossRef]
44. Fukue, M.; Yanai, M.; Sato, Y.; Fujikawa, T.; Furukawa, Y.; Tani, S. Background Values for Evaluation of Heavy Metal Contamination in Sediments. *J. Hazard. Mater.* **2006**, *136*, 111–119. [CrossRef]
45. Maerki, M.; Wehrli, B.; Dinkel, C.; Müller, B. The Influence of Tortuosity on Molecular Diffusion in Freshwater Sediments of High Porosity. *Geochim. Cosmochim. Acta* **2004**, *68*, 1519–1528. [CrossRef]
46. Shen, L.; Chen, Z. Critical Review of the Impact of Tortuosity on Diffusion. *Chem. Eng. Sci.* **2007**, *62*, 3748–3755. [CrossRef]

toxics

MDPI

Article

Association of Urinary Cadmium and Antimony with Osteoporosis Risk in Postmenopausal Brazilian Women: Insights from a 20 Metal(loid) Biomonitoring Study

Carlos Tadashi Kunioka [1,2], Vanessa Cristina de Oliveira Souza [3], Bruno Alves Rocha [3,†], Fernando Barbosa Júnior [3], Luís Belo [4], Maria Conceição Manso [1,5,6] and Márcia Carvalho [1,5,6,*]

1 FP-I3ID, FP-BHS, Fernando Pessoa University, Praça de 9 de Abril 349, 4249-004 Porto, Portugal; 36181@ufp.edu.pt (C.T.K.); cmanso@ufp.edu.pt (M.C.M.)
2 Western Paraná State University (UNIOESTE), Cascavel 85819-110, PR, Brazil
3 Department of Clinical Analyses, Toxicology and Food Sciences, School of Pharmaceutical Sciences of Ribeirao Preto, University of Sao Paulo, Ribeirão Preto 14040-903, SP, Brazil; vcosouza@fcfrp.usp.br (V.C.d.O.S.); farmbrunorocha@gmail.com (B.A.R.); fbarbosa@fcfrp.usp.br (F.B.J.)
4 UCIBIO i4HB, Faculty of Pharmacy, University of Porto, Rua Jorge Viterbo Ferreira 228, 4050-313 Porto, Portugal; luisbelo@ff.up.pt
5 LAQV/REQUIMTE, University of Porto, 4050-313 Porto, Portugal
6 RISE-Health, Faculty of Health Sciences, Fernando Pessoa University, Fernando Pessoa Teaching and Culture Foundation, Rua Carlos da Maia 296, 4200-150 Porto, Portugal
* Correspondence: mcarv@ufp.edu.pt
† Current address: Institute of Chemistry, Federal University of Alfenas, Alfenas 37130-001, MG, Brazil.

Abstract: Osteoporosis is a major public health concern, particularly among postmenopausal women. Environmental exposure to metals has been proposed as a potential contributor to osteoporosis, but human data remain limited and inconsistent. This study investigated changes in urinary concentrations of 20 metal(loid)s in patients with osteoporosis, as well as the association of these elements with bone mineral density (BMD), in a cohort of 380 postmenopausal women aged 50–70 years from Cascavel, Paraná, Brazil. Demographic, lifestyle, and clinical data were collected, and urinary concentrations of aluminum (Al), barium (Ba), cadmium (Cd), cobalt (Co), cesium (Cs), copper (Cu), mercury (Hg), lithium (Li), manganese (Mn), molybdenum (Mo), nickel (Ni), lead (Pb), rubidium (Rb), antimony (Sb), selenium (Se), tin (Sn), strontium (Sr), thallium (Tl), uranium (U), and zinc (Zn) were measured by inductively coupled plasma mass spectrometry. BMD was assessed at the lumbar spine, femoral neck, and total hip using dual-energy X-ray absorptiometry. Osteoporosis was diagnosed in 73 participants (19.2%). Osteoporotic women had significantly higher urinary concentrations of Cd, Mn, Pb, Sb, Sn, and Zn ($p < 0.05$). Statistically significant negative correlations were observed between BMD and urinary concentrations of Al, Cd, Hg, Mn, Sb, and U. After adjustment for confounders, elevated urinary concentrations of Cd, Mn, Pb, and Sb remained independently and significantly associated with higher odds of osteoporosis, with Cd (aOR = 1.495; $p = 0.026$) and Sb (aOR = 2.059; $p = 0.030$) showing the strongest associations. In addition, women with urinary concentrations above the 90th percentile for both Cd and Sb had a significantly higher prevalence of osteoporosis compared to those with lower levels (44.4% vs. 18.0%; $p = 0.011$). Longitudinal studies are needed to confirm causality and inform prevention strategies.

Keywords: metals; metalloids; environmental exposure; bone mineral density; women; aging; osteoporosis

1. Introduction

Osteoporosis is a highly prevalent osteometabolic disorder characterized by a reduction in bone mass and the deterioration of bone microarchitecture, resulting in increased bone fragility and risk of fracture [1,2]. This condition is recognized as a major public health concern worldwide, particularly among postmenopausal women [3], driven by the dual impact of estrogen deficiency and aging on bone health [4]. Estrogen plays a key role in regulating bone remodeling by inhibiting osteoclast-mediated bone resorption and supporting osteoblast function. Following menopause, the abrupt decline in estrogen levels disrupts this balance, resulting in accelerated bone loss [4,5]. Concurrently, aging contributes to a gradual decline in bone formation due to reduced osteoblast activity, impaired calcium absorption, and changes in bone microarchitecture [6,7]. This combination of hormonal and age-related factors synergistically increases the risk of osteoporotic fractures, especially those affecting the hip and spine, leading to significant morbidity, long-term disability, reduced quality of life, and increased mortality [8].

Osteoporosis affects an estimated 500 million people worldwide, with women facing a lifetime risk of osteoporotic fractures between 30–50%, compared to 15–30% in men [9]. According to the Global Burden of Disease Study 2019, low BMD was responsible for 438,000 deaths and 16.6 million disability-adjusted life years (DALYs) worldwide [1]. That same year, 178 million new fractures were recorded, contributing to 25.8 million years lived with disability (YLDs) [10]. As global populations continue to age, the global burden of osteoporosis is projected to rise sharply, creating significant challenges to healthcare systems. This issue is further compounded by the feminization of aging, as, on average, women live longer than men and are disproportionately affected by the disease.

In Brazil, osteoporosis also poses a significant and growing public health challenge. Recent national surveys estimate the prevalence of osteoporosis among Brazilian women over the age of 50 to range between 15% and 33%, depending on the diagnostic criteria and region studied [11]. The aging demographic profile of the Brazilian population, combined with lifestyle factors such as sedentary behavior, sub-optimal nutrition, and low vitamin D concentrations, has contributed to an increasing incidence of osteoporotic fractures. Hip fractures, in particular, have shown a rising trend in hospitalization rates, imposing significant costs on the Brazilian Unified Health System (SUS) and affecting patients' functional independence and quality of life [8].

Given its significant clinical and socioeconomic impact, understanding the modifiable risk factors for osteoporosis is essential to guide public health interventions and preventive strategies. While traditional determinants, including age, sex, hormonal status, calcium and vitamin D intake, and physical activity, are well established [12], growing evidence suggests that environmental exposures to toxic metals, such as lead, cadmium, and arsenic, may also play a significant role in bone health and the development of osteoporosis [13–15], and this warrants further investigation.

Metals and metalloids are ubiquitous in the environment due to both natural and anthropogenic activities, and humans can be exposed through inhalation, ingestion, or dermal contact [16,17]. Several studies have investigated the influence of essential and toxic elements on bone health, revealing complex and sometimes contradictory effects. Essential trace elements, such as copper (Cu), manganese (Mn), selenium (Se), zinc (Zn), and cobalt (Co), are essential for maintaining normal physiological functions, including bone metabolism [18]. However, studies have shown that high levels of manganese (Mn) exposure may be associated with an increased risk of osteoporosis [19]. Zn has been shown to promote osteoblast proliferation and differentiation and bone matrix formation [20]. Selenium (Se), recognized for its antioxidant properties, may protect against oxidative stress [21,22], a factor that has been implicated in bone loss, and cobalt (Co) is involved in vitamin B12 metabolism, which may indirectly influence bone health [23]. However, both deficient and excessive intakes of these essential metals can have adverse effects [18,23].

Numerous epidemiological studies have highlighted the detrimental effects of exposure to toxic metals, particularly cadmium (Cd) and lead (Pb), on BMD and fracture risk [24–26]. Cd has been consistently linked to adverse skeletal outcomes. In a recent systematic review and meta-analysis conducted by our group, we demonstrated that even low-level environmental Cd exposure is linked to an increased risk of osteoporosis in postmenopausal women [27]. Similarly, low-level Pb exposure has been implicated in bone demineralization and disturbances in calcium homeostasis [28]. The effects of mercury (Hg) on bone health are less well defined. Although human data are lacking, some studies link Hg exposure to altered calcium homeostasis and increased osteoclast activity [29], while others report inconsistent findings [30]. Furthermore, elements such as antimony (Sb) and thallium (Tl) have been shown to induce oxidative stress [31–33], a process detrimental to bone integrity. However, their specific roles and underlying mechanisms in bone metabolism remain poorly understood and warrant further investigation.

Other less common elements, such as lithium (Li) and strontium (Sr), present more complex or even contradictory effects, depending on the exposure level and the population studied [34]. Lithium has been suggested to promote bone formation through the activation of the Wnt/β-catenin signaling pathway [35,36], a critical signaling pathway in osteogenesis [37], while Sr, although used therapeutically in osteoporosis management, may impair normal bone mineralization when present at high environmental levels [18,38,39]. Uranium (U) and cesium (Cs) are radioactive elements that may also interfere with calcium homeostasis in bone [40–42], though their effects in human populations remain poorly understood.

This cohort study seeks to address existing knowledge gaps by investigating the associations between exposure to a broad range of metals and metalloids—including aluminum (Al), barium (Ba), Cd, Co, Cs, Cu, Hg, Li, Mn, molybdenum (Mo), nickel (Ni), Pb, rubidium (Rb), Sb, Se, tin (Sn), Sr, Tl, U, and Zn—and BMD in postmenopausal women. By biomonitoring urinary metal(loid) levels, this study offers novel insights into how environmental exposure to these metals may affect bone health and contribute to osteoporosis risk.

2. Methods

2.1. Study Design and Population

We conducted a cross-sectional study using a cohort of 380 postmenopausal women. Ethical approval for the study was granted by the Research Ethics Committees of the State University of Western Paraná (Unioeste), Plataforma Brasil (approval number: 2.636.746), and Fernando Pessoa University in Porto, Portugal. Prior to participation, all individuals provided written informed consent.

Data collection spanned from March 2022 to February 2024. Participants were recruited through announcements disseminated via social media platforms, universities, medical clinics, health centers, hospitals, and both regional and municipal health departments. The inclusion criteria specified women between 50 and 70 years of age who were postmenopausal, defined retrospectively as experiencing at least 12 consecutive months without menstruation, and who had been residing in the study region for a minimum of 10 years. Exclusion criteria included the presence of serious active medical conditions such as advanced hepatic or renal insufficiency and cancer, a medically confirmed diagnosis of secondary osteoporosis—including but not limited to hyperparathyroidism, malignancies, or chronic corticosteroid therapy—as well as any history of occupational exposure to metals through industrial employment.

After obtaining informed consent, participants completed a structured questionnaire designed to collect data on osteoporosis risk factors, potential dietary and occupational metal exposures, as well as relevant sociodemographic and health-related variables. Prolonged bed rest was defined as a duration of 28 days or more, while insufficient physical activity was classified as engaging in less than 30 min of exercise per day. Following the questionnaire, trained research staff measured each participant's body weight and height, from which body mass index (BMI) was calculated. The research team then coordinated the scheduling of urine sample collection and bone mineral density (BMD) assessments for all participants.

2.2. BMD Measurement

Bone mineral density (BMD) assessment was conducted for all participants using dual-energy X-ray absorptiometry (DEXA). The examinations were performed within the same radiology department utilizing the HOLOGIC Horizon-A model device. Measurements were obtained at standardized anatomical locations, specifically the lumbar spine, femoral neck, and total hip. The classification of bone health status followed the guidelines established by the World Health Organization (WHO). According to these criteria, osteoporosis is diagnosed when BMD is reduced by 2.5 SD or more at any of the evaluated bone sites.

2.3. Measurement of Urinary Metal(loid) Concentrations

First-morning urine samples were collected in metal-free tubes and stored at −20 °C until analysis. Urinary metal and metalloid concentrations were performed using inductively coupled plasma mass spectrometry (ICP-MS), equipped with a quadrupole ion deflector (NexION® 2000, PerkinElmer, Shelton, CT, USA) and operated with high-purity argon gas (99.999%, Air Liquide, São Paulo, Brazil), following the analytical protocol previously established in the laboratory [43]. All reagents employed were of analytical grade (Sigma-Aldrich, St. Louis, MO, USA). Nitric acid (HNO_3) used in sample preparation was further purified through sub-boiling distillation in a quartz distillation apparatus (Kürner Analysentechnik).

For each element analyzed, calibration curves with matrix matching were constructed using standard solutions ranging from 0 to 200 μg/L, prepared with a diluent composed of 0.5% HNO_3 and 0.01% Triton X-100. Method accuracy and precision were evaluated using certified reference urine samples provided by the Institut National de Santé Publique du Québec (INSPQ) (QM-U-Q1509, Quebec, Canada). A 200 μL aliquot of each urine sample was diluted to a final volume of 5 mL with the prepared diluent, and all measurements were conducted in triplicate. Urinary element concentrations falling below the limit of detection (LOD) were estimated by assigning a value equal to the LOD divided by the square root of two ($LOD/\sqrt{2}$).

Urinary creatinine concentrations were determined using the alkaline picrate method on the Atellica® CH analyzer (Siemens, São Paulo, Brazil). Metal and metalloid concentrations were then normalized to creatinine concentrations in urine and expressed as μg/g creatinine.

All procedures were performed at the Analytical and Systems Toxicology Laboratory (ASTox), Ribeirão Preto School of Pharmaceutical Sciences, University of São Paulo.

2.4. Statistical Analysis

Categorical variables were summarized as absolute and relative frequencies (n, %) and compared using the chi-square test or Fisher's exact test, as appropriate. Quantitative variables were presented as medians with interquartile ranges (IQRs; Q1–Q3) and compared between groups using the Mann–Whitney U test due to non-normal distributions.

Associations between key variables were assessed using Spearman's rank or Pearson correlation coefficients, with ln-transformation applied when necessary. A heatmap based on Spearman's correlation values (selected due to non-normality) was used to visualize relationships among urinary concentrations of the analyzed elements.

Multiple linear regression with stepwise selection was used to identify independent variables associated with BMD; variables were ln-transformed when needed to meet model assumptions. Additionally, binary logistic regression was performed to evaluate the association between potential predictors and osteoporosis risk. Odds ratios (ORs) were adjusted for age, BMI, duration of menopause, smoking status, and prolonged bed rest.

All analyses were performed using IBM SPSS Statistics version 30.0. A two-tailed p-value < 0.05 was considered statistically significant.

3. Results

3.1. Characteristics of the Study Population

The sociodemographic characteristics, health habits, and clinical data of the postmenopausal women included in this study are summarized in Table 1. The median age of the 380 participants was 60 years. The median BMI was 27 kg/m^2, with 75% of the women presenting a BMI above the upper-normal cutoff of 25 kg/m^2. The majority of participants (93.7%) reported never having smoked or being former smokers, and only one woman reported current alcohol consumption. The overall prevalence of osteoporosis in the study population was 19.2%.

Table 1. Clinical and sociodemographic characteristics of participants, presented overall and according to osteoporosis diagnosis.

Variables	Overall n = 380	No Osteoporosis n = 307	With Osteoporosis n = 73	p-Value [1]
Age (years)	60.0 (56.0; 65.8)	60.0 (56.0; 65.0)	62.0 (59.0; 66.0)	**0.011**
BMI (kg/m^2)	27.0 (24.4; 30.2)	27.6 (24.8; 30.8)	26.1 (23.4; 28.1)	**<0.001**
Length of menopause (years)	13.0 (7.3; 19.0)	12.0 (7.0; 18.0)	16.0 (10.5; 21.5)	**0.002**
Prior fracture (yes)	109 (28.7%)	84 (27.4%)	25 (34.2%)	0.252
Arthritis (yes)	72 (18.9%)	62 (20.2%)	10 (13.7%)	0.246
Vitamin D intake (yes)	183 (48.2%)	149 (48.5%)	34 (46.4%)	0.795
Corticoids (yes)	88 (23.2%)	70 (22.8%)	18 (24.7%)	0.758
Prolonged bed rest (yes)	28 (7.4%)	21 (6.8%)	7 (9.6%)	0.454
No exercise (yes)	135 (35.5%)	112 (36.5%)	23 (31.5%)	0.497
Calcium intake (yes)	100 (26.3%)	68 (22.1%)	32 (43.8%)	**<0.001**
Alcohol intake (yes)	1 (0.3%)	1 (0.3%)	0 (0%)	1.000
Smoking (yes)	24 (6.3%)	16 (5.2%)	8 (11.0%)	0.333
Antiresorptive medications				**<0.001**
Bisphosphonates	10 (2.6%)	4 (1.3%)	6 (8.2%)	
Bisphosphonates and HRT	1 (0.3%)	0 (0%)	1 (1.4%)	
HRT	45 (11.8%)	33 (10.7%)	12 (16.4%)	
No	324 (85.3%)	270 (87.9%)	54 (74.0%)	
Lumbar spine				
BMD (g/cm^2)	0.92 (0.82; 1.06)	0.97 (0.88; 1.08)	0.74 (0.70; 0.77)	**<0.001**
T-score	−1.10 (−2.00; 0.10)	−0.70 (−1.50; 0.40)	−2.80 (−3.10; −2.50)	**<0.001**
Diagnosis of osteoporosis	61 (16.1%)	0 (0%)	61 (83.6%)	**<0.001**
Femoral neck				
BMD (g/cm^2)	0.73 (0.65; 0.83)	0.75 (0.68; 0.86)	0.63 (0.56; 0.71)	**<0.001**

Table 1. *Cont.*

Variables	Overall n = 380	No Osteoporosis n = 307	With Osteoporosis n = 73	*p*-Value [1]
T-score	−1.10 (−1.78; −0.20)	−0.90 (−1.50; 0.10)	−2.00 (−2.65; −1.35)	**<0.001**
Diagnosis of osteoporosis	25 (6.6%)	0 (0%)	25 (34.2%)	**<0.001**
Total hip				
BMD (g/cm^2)	0.87 (0.78; 0.96)	0.90 (0.81; 0.98)	0.74 (0.69; 0.84)	**<0.001**
T-score	−0.60 (−1.30; 0.10)	−0.40 (−1.00; 0.30)	−1.60 (−2.05; −0.90)	**<0.001**
Diagnosis of osteoporosis	10 (2.6%)	0 (0%)	10 (13.7%)	**<0.001**
Urinary creatinine (mg/dL)	52.7 (29.0; 90.9)	52.8 (31.1; 93.1)	52.5 (22.2; 85.4)	0.090

Results are presented as median (first quartile; third quartile), unless otherwise indicated; [1] Mann–Whitney test; Statistically significant differences are shown in bold. Abbreviations: BMD, Bone mineral density; BMI, Body mass index; HRT, Hormone replacement therapy. Statistically significant *p*-values are shown in bold.

Table 1 also shows a comparison of these variables between women with and without a diagnosis of osteoporosis. Statistically significant differences were observed between the two groups in terms of age, BMI, menopausal duration, and calcium intake. Specifically, women diagnosed with osteoporosis were slightly older, had a lower BMI, experienced a longer postmenopausal period, and reported higher calcium intake. Additionally, a significantly higher proportion of women with osteoporosis reported using antiresorptive medication ($p < 0.001$).

As expected, bone mineral density (BMD) and T-scores at the lumbar spine, femoral neck, and total hip were significantly lower in women with osteoporosis compared to those without osteoporosis ($p < 0.001$).

3.2. Urinary Concentrations of Metals and Metalloids in Brazilian Postmenopausal Women

Table 2 depicts the urinary concentrations of 20 metals and metalloids for the entire study cohort, as well as separately for women with and without an osteoporosis diagnosis. After adjusting for age, BMI, duration of menopause, smoking status, and prolonged bed rest, statistically significant differences in median urinary concentrations were identified between the two groups for Cd, Mn, Pb, Sb, Sn, and Zn. Specifically, women with osteoporosis had higher median urinary concentrations of Cd (0.38 µg/g creatinine vs. 0.30 µg/g creatinine, $p = 0.012$), Mn (7.0 µg/g creatinine vs. 4.1 µg/g creatinine, $p = 0.014$), Pb (4.8 µg/g creatinine vs. 3.7 µg/g creatinine, $p = 0.020$), Sb (0.27 µg/g creatinine vs. 0.17 µg/g creatinine, $p = 0.015$), Sn (0.70 µg/g creatinine vs. 0.49 µg/g creatinine, $p = 0.046$), and Zn (860 µg/g creatinine vs. 777 µg/g creatinine, $p = 0.004$) compared to women without osteoporosis.

Table 2. Urinary concentrations (expressed in µg/g creatinine) of metals and metalloids, presented overall and according to osteoporosis diagnosis.

Elements (µg/g creat)	Overall n = 380	No Osteoporosis n = 307	With Osteoporosis n = 73	*p*-Value [1]
Al	217 (108; 448)	202 (106; 425)	340 (129; 638)	0.273
Ba	16.5 (8.3; 29.0)	15.4 (8.0; 27.8)	19.7 (9.0; 36.8)	0.127
Cd	0.30 (0.15; 0.55)	0.30 (0.14; 0.49)	0.38 (0.16; 0.71)	**0.012**
Co	0.23 (0.09; 0.46)	0.22 (0.08; 0.46)	0.26 (0.13; 0.66)	0.543
Cs	7.9 (4.0; 14.0)	7.7 (4.0; 13.5)	8.2 (4.6; 15.5)	0.419
Cu	93.6 (53.1; 172.9)	92.0 (52.9; 169.9)	98.4 (57.6; 184.7)	0.148
Hg	0.99 (0.41; 2.07)	0.97 (0.40; 1.93)	1.03 (0.55; 2.91)	0.501
Li	6.9 (3.2; 14.9)	6.7 (3.2; 14.8)	8.5 (3.9; 17.0)	0.578

Table 2. *Cont.*

Elements (μg/g creat)	Overall n = 380	No Osteoporosis n = 307	With Osteoporosis n = 73	*p*-Value [1]
Mn	4.4 (1.7; 9.7)	4.1 (1.7; 8.8)	7.0 (2.4; 15.7)	**0.014**
Mo	17.0 (8.6; 33.4)	17.0 (8.6; 32.3)	16.5 (9.3; 37.7)	0.378
Ni	16.0 (8.1; 30.9)	15.8 (8.0; 28.9)	17.5 (9.1; 35.9)	0.219
Pb	4.0 (2.1; 8.1)	3.7 (2.1; 7.4)	4.8 (2.7; 11.7)	**0.020**
Rb	3105 (1567; 5742)	3026 (1557; 5627)	3636 (1649; 6220)	0.570
Sb	0.19 (0.10; 0.39)	0.17 (0.10; 0.36)	0.27 (0.13; 0.52)	**0.015**
Se	11.7 (5.9; 24.4)	12.0 (6.0; 25.6)	10.8 (5.5; 22.9)	0.439
Sn	0.52 (0.23; 1.09)	0.49 (0.22; 0.92)	0.70 (0.24; 1.86)	**0.046**
Sr	76.4 (40.7; 179.3)	75.8 (38.5; 178.7)	87.9 (51.3; 215.3)	0.100
Tl	0.22 (0.11; 0.46)	0.21 (0.11; 0.45)	0.26 (0.12; 0.54)	0.076
U	0.017 (0.005; 0.041)	0.016 (0.005; 0.038)	0.021 (0.008; 0.056)	0.109
Zn	808 (458; 1548)	777 (460; 1510)	860 (456; 1882)	**0.004**

Results are presented as median (first quartile; third quartile); [1] Generalized Linear Model (GLM) of the ln-transformed elements adjusted for age (years), BMI (Kg/m^2), length of menopause (years), smoking, and prolonged bed rest. Statistically significant differences are shown in bold. Abbreviations: Al, aluminum; Ba, barium; Cd, cadmium; Co, cobalt; Cs, cesium; Cu, copper; Hg, mercury; Li, lithium; Mn, manganese; Mo, molybdenum; Ni, nickel; Pb, lead; Rb, rubidium; Sb, antimony; Se, selenium; Sn, tin; Sr, strontium; Tl, thallium; U, uranium; Zn, zinc. Statistically significant *p*-values are shown in bold.

3.3. Intercorrelations Between Metal(loid) Concentrations in Urine

Figure 1 presents the heatmap generated from Spearman correlation coefficients calculated for all measured metal(loid) concentrations in urine of the study population. Notably, several strong positive correlations were observed between creatinine-adjusted element concentrations. The strongest correlations in decreasing order were between Rb and Cs (0.955, $p < 0.01$), Tl and Rb (0.904, $p < 0.01$), Se and Cs (0.892, $p < 0.01$), Tl and Cs (0.890, $p < 0.01$), Mn and Al (0.871, $p < 0.01$), Ni and Ba (0.869, $p < 0.01$), Se and Rb (0.866, $p < 0.01$), Zn and Ni (0.858, $p < 0.01$), Mo and Cs (0.850, $p < 0.01$), Ni and Cu (0.849, $p < 0.01$), Zn and Cu (0.847, $p < 0.01$), Cu and Cs (0.845, $p < 0.01$), Zn and Ba (0.843, $p < 0.01$), Ba and Al (0.840, $p < 0.01$), Rb and Mo (0.832, $p < 0.01$), Zn and Cu (0.828, $p < 0.01$), Cu and Ba (0.826, $p < 0.01$), Sr and Cs (0.820, $p < 0.01$), Rb and Cu (0.816, $p < 0.01$), Pb and Ba (0.814, $p < 0.01$), Sb and Ni (0.811, $p < 0.01$), Pb and Al (0.808, $p < 0.01$), and Sb and Ba (0.805, $p < 0.01$).

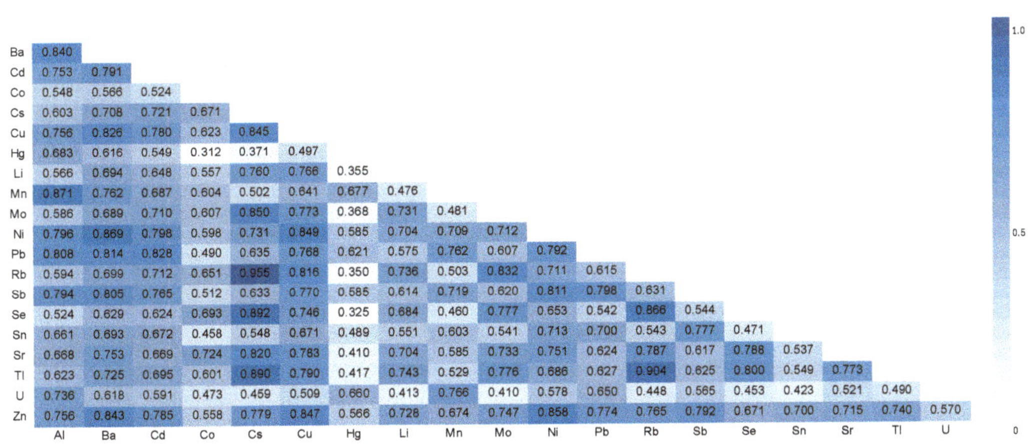

Figure 1. Heatmap plot showing correlation between the elements in urine. Spearman's rank correlation coefficients (R).

3.4. Association of Studied Variables (Clinical and Analytical) with BMD

The statistically significant associations obtained between BMD at the lumbar spine, femoral neck, and total hip and age, BMI, menopause length, and urinary metal(loid) concentrations are presented in Table 3. Age and menopause length showed negative correlations with BMD at all three sites, while BMI showed positive correlations. Among the metals and metalloids, weak but statistically significant negative correlations were observed between BMD and urinary concentrations of Al (lumbar spine, total hip), Cd (lumbar spine, total hip), Hg (total hip), Mn (lumbar spine), Sb (lumbar spine, total hip), and U (lumbar spine, femoral neck, total hip).

Table 3. Statistically significant associations between ln-transformed bone mineral density (BMD) at the lumbar spine, femoral neck, and total hip and ln-transformed age, BMI, menopause length, and urinary metal(loid) concentrations.

		BMD Lumbar Spine	BMD Femoral Neck	BMD Total Hip
BMD Lumbar Spine	r	1	0.525	0.697
	p		**<0.001**	**<0.001**
BMD Femoral Neck	r	0.525	1	0.743
	p	**<0.001**		**<0.001**
BMD Total Hip	r	0.697	0.743	1
	p	**<0.001**	**<0.001**	
Age	r	−0.182	−0.256	−0.278
	p	**<0.001**	**<0.001**	**<0.001**
BMI	r	0.301	0.311	0.456
	p	**<0.001**	**<0.001**	**<0.001**
Menopause Length	r	−0.247	−0.220	−0.271
	p	**<0.001**	**<0.001**	**<0.001**
Al	r	−0.112	−0.062	−0.104
	p	**0.030**	0.227	**0.042**
Cd	r	−0.102	−0.064	−0.128
	p	**0.048**	0.215	**0.013**
Hg	r	−0.084	−0.062	−0.128
	p	0.101	0.226	**0.012**
Mn	r	−0.128	−0.072	−0.097
	p	**0.012**	0.159	0.058
Sb	r	−0.106	−0.068	−0.104
	p	**0.039**	0.186	**0.044**
U	r	−0.107	−0.103	−0.116
	p	**0.036**	**0.044**	**0.023**

r, Pearson correlation coefficient; Abbreviations: Al, aluminum; BMD, bone mineral density; BMI, body mass index; Cd, cadmium; Hg, mercury; Mn, manganese; Sb, antimony; U, uranium. Statistically significant *p*-values are shown in bold.

Given that lumbar spine was the bone site where most cases of osteoporosis were diagnosed, we sought to determine the primary determinants of BMD at this location. Multiple linear regression analysis identified ln-transformed BMI as the strongest positive predictor of lumbar spine BMD, while ln-transformed menopause length, smoking, and prolonged bed rest were significant negative predictors (Table 4). No metals or metalloids were included as significant predictors of lumbar spine BMD in this model.

Table 4. Main variables associated with bone mineral density at the lumbar spine, by multiple linear regression analysis.

Dependent Variable	Model	Unstandardized Coefficients		Standardized Coefficients	t	*p*-Value
		Beta (95% CI)	Std. Error	Beta		
Ln BMD	(Constant)	−0.956 (−1.277; −0.636)	0.163		−5.863	**<0.001**
	Ln BMI	0.310 (0.214; 0.405)	0.049	0.299	6.366	**<0.001**
	Ln menopause Length	−0.052 (−0.072; −0.032)	0.010	−0.242	−5.140	**<0.001**
	Smoking	−0.024 (−0.045; −0.004)	0.010	−0.111	−2.358	**0.019**
	Prolonged bed rest	−0.067 (−0.129; −0.004)	0.032	−0.099	−2.096	**0.037**

Abbreviations: BMD, bone mineral density; BMI, body mass index. Statistically significant *p*-values are shown in bold.

3.5. Association Between Urinary Metal(loid) Concentrations and Osteoporosis Risk

Independent adjusted odds ratios (aORs) for osteoporosis outcome associated with ln-transformed urinary metal and metalloid variables are presented in Figure 2. After adjusting for age, BMI, length of menopause, smoking, and prolonged bed rest, urinary Cd (aOR = 1.495, 95% CI: 1.048; 2.131, $p = 0.026$), Mn (aOR = 1.014, 95% CI: 1.001; 1.028, $p = 0.040$), Sb (aOR = 2.059, 95% CI: 1.073; 3.950, $p = 0.030$), and Zn (aOR = 1.00027, 95% CI: 1.00006; 1.00048, $p = 0.012$) were significantly associated with osteoporosis outcome. However, only Cd and Sb showed clinically meaningful associations, as the odds ratios for Mn and Zn were very close to 1.

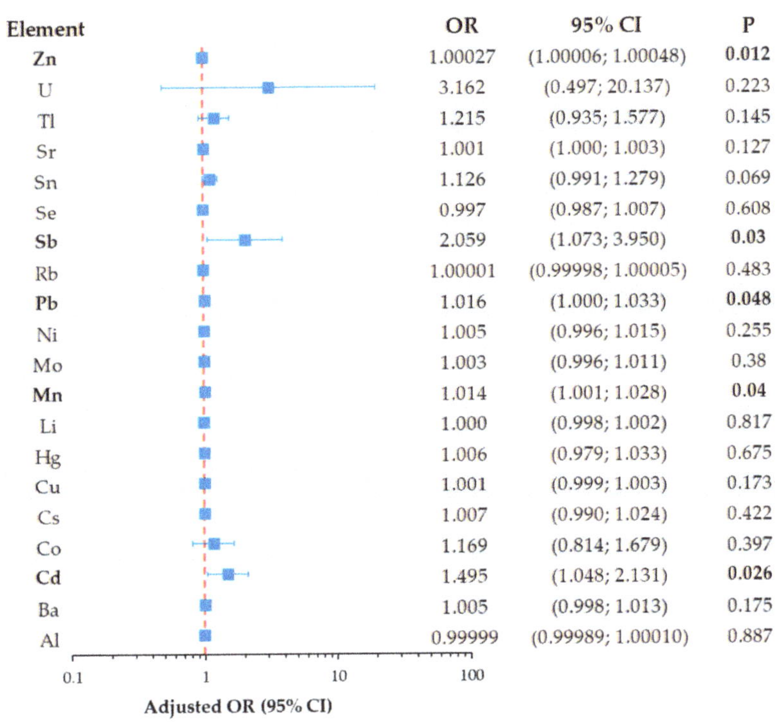

Figure 2. Forest plot showing the independent adjusted odds ratios (aORs) for osteoporosis outcomes associated with metal(loid) exposure. ORs were adjusted for age, BMI, length of menopause, smoking, and prolonged bed rest. Statistically significant *p*-values are shown in bold.

Given that Cd and Sb demonstrated the strongest and most clinically meaningful associations with increased osteoporosis risk, we further analyzed their combined effects. Women were categorized based on whether their urinary concentrations of both elements were below or at/above the 90th percentile thresholds (1.079 μg/g creatinine for cadmium and 0.729 μg/g creatinine for antimony; see Supplementary Table S1). Notably, women with elevated urinary concentrations exhibited significantly lower BMD at all measured bone sites and a higher prevalence of osteoporosis (44.4% vs. 18.0%; $p = 0.011$).

4. Discussion

This cross-sectional study investigated the association between urinary concentrations of 20 metal(loid)s and osteoporosis in a cohort of postmenopausal Brazilian women. Several significant associations were identified. Consistent with established risk factors for osteoporosis, women diagnosed with the disease were generally older, had a lower BMI, and experienced a longer time since menopause onset. Moreover, increased urinary concentrations of aluminum, cadmium, mercury, manganese, antimony, and uranium were correlated with decreased BMD. Notably, this study is the first to demonstrate a significant association between elevated urinary concentrations of both cadmium and antimony and an increased risk of postmenopausal osteoporosis, even after adjusting for potential confounders. In light of the increasing environmental burden of these elements from industrial, mining, and vehicular sources [44], our findings reinforce the importance of further human and environmental monitoring studies.

Biomonitoring studies are crucial for assessing population-level exposure to environmental metals and metalloids, identifying vulnerable groups, and informing risk mitigation efforts. Beyond characterizing exposure–outcome relationships, biomonitoring provides an essential tool for evaluating the effectiveness of regulatory policies over time [45,46]. Urine is widely recognized as a suitable biological matrix for such assessments, owing to its non-invasive collection and its reliability for specific metals. For elements such as cadmium, cobalt, cesium, molybdenum, nickel, antimony, strontium, thallium, and uranium, urinary concentrations serve as robust biomarkers of exposure due to their relatively stable excretion profiles [47]. In contrast, for metals such as lead, mercury, manganese, aluminum, copper, and zinc, urinary levels are considered less reliable indicators of exposure [48]. This limitation arises from alternative primary excretion routes (e.g., biliary or fecal pathways) or tight homeostatic regulation, which can obscure the relationship between exposure and urinary elimination.

In our study, urinary concentrations of two essential trace metals—manganese and zinc—and of four toxic metal(loid)s—cadmium, lead, tin, and antimony—were higher in osteoporotic women compared with non-osteoporotic controls (Table 2). Both Mn and Zn were reported to present a U-shaped curve in terms of their toxicity and effect on bone health. This means that, at low levels, they are essential and beneficial for bone formation and mineralization, whereas, at excessively high levels, they become toxic, leading to impaired bone metabolism and bone loss [18]. Therefore, the increased urinary excretion of Mn and Zn observed in the osteoporosis group may reflect heightened environmental exposure but could also result from osteoporosis-induced loss of essential trace elements, as previously described [49]. This loss may further exacerbate deficiencies that are critical for maintaining bone health. Conversely, the elevated urinary concentrations of Cd, Pb, Sn, and Sb are more likely attributable to increased environmental exposure to these toxic metals, which are known to interfere with bone remodeling and mineralization, thereby potentially contributing to osteoporosis risk [14]. Among them, in the present study only cadmium and antimony showed significant negative correlations with BMD (Table 3), implicating a potential role in bone deterioration. However, the observed correlation coefficients were modest, indicating weak predictive value when considered individually. This was corroborated by multiple linear regression analysis, which failed to identify any of the metals as independent predictors of lumbar spine BMD (Table 4).

Nevertheless, potential adverse effects may become more pronounced at higher exposure levels, as we recently demonstrated for Cd in Brazilian postmenopausal women [50], or in scenarios involving co-exposure to multiple metal(loid)s. In our study, women with urinary cadmium and antimony concentrations above the 90th percentile exhibited a significantly higher prevalence of postmenopausal osteoporosis (Supplementary Table S1). This finding underscores the possible additive or synergistic effects of Cd and Sb on bone degradation, in agreement with previous reports on the combined impact of metal mixtures on BMD reduction [51]. Notably, both cadmium and antimony tend to accumulate in soil and crops due to industrial emissions and agricultural practices [52,53], increasing the likelihood of concurrent human exposure through dietary intake or shared environmental sources. This co-exposure scenario is further supported by the strong positive correlation observed between urinary concentrations of these metals in our cohort ($p = 0.765$; Figure 1).

Substantial epidemiological evidence indicates that chronic exposure to cadmium is associated with impaired bone metabolism, reduced BMD, and increased risk of osteopenia and osteoporosis [14,27,54,55]. Cadmium exerts its deleterious effects through multiple mechanisms, including renal dysfunction, disruption of calcium homeostasis, oxidative stress, and inflammation. Specifically, cadmium inhibits osteoblast differentiation while enhancing osteoclast activity, partly through interference with Wnt/β-catenin signaling and upregulation of RANKL expression [15,56]. Cadmium-induced oxidative stress activates NF-κB and increases proinflammatory cytokines such as IL-6 and TNF-α, further contributing to bone resorption [57]. Additionally, cadmium disrupts calcium metabolism by impairing gastrointestinal absorption, renal reabsorption, and vitamin D receptor-mediated synthesis of 1,25-dihydroxyvitamin D [58]. Proteomic and metabolomic analyses have also revealed cadmium-related alterations in osteogenic gene expression, bone matrix protein production, and mineralization pathways, further implicating cadmium in skeletal fragility [59]. In contrast, data on antimony remain limited. Given its chemical similarity to arsenic, a recognized osteotoxicant, antimony may exert analogous effects, including oxidative stress induction and disruption of osteoblast function [33]. Some experimental studies have linked high antimony levels to redox imbalance and increased bone loss [60]. However, the evidence remains sparse and inconsistent, highlighting the need for additional in vivo and in vitro studies to elucidate antimony's role in bone demineralization and osteoporosis risk.

The adjusted odds ratios provided deeper insight into the independent relationships between metals and metalloid exposure and osteoporosis risk (Figure 2). Antimony demonstrated the strongest association (aOR = 2.059), suggesting that higher urinary antimony concentrations are associated with more than a two-fold increase in the odds of developing osteoporosis. Cadmium was also associated with an increased risk (aOR = 1.495), indicating a moderate but clinically meaningful effect. Statistically significant associations were observed for manganese and zinc as well; however, their effect sizes were minimal, as indicated by the low odds ratios. The significance for manganese and zinc likely reflects the large sample size and narrow confidence intervals rather than a meaningful biological impact.

To the best of our knowledge, only two studies to date have investigated the associations between urinary concentrations of multiple metallic elements and BMD loss or osteoporosis risk in older and/or postmenopausal women. The first study, based on data from the U.S. National Health and Nutrition Examination Survey (NHANES) 2005–2010, supports our findings by identifying urinary cadmium, along with arsenic and tungsten, as negatively associated with BMD [61]. A more recent analysis of NHANES 2017–2020 data found no statistically significant associations between BMD in women and any of the 11 examined metal(loid)s (Ba, Cd, Co, Cs, Mo, Mn, Pb, Sb, Sn, Tl, and W) after adjusting for potential confounders. However, a consistent negative trend was observed for all elements except manganese. This may reflect the impact of successful public health interventions that have substantially reduced environmental exposures in the U.S. population over time [62]. As a result, current exposure levels may be below the biological threshold needed to influence bone health. Notably, the NHANES 2017–2020 study reported a sex-specific effect of antimony: urinary antimony levels were negatively associated with femoral BMD in women but positively associated in men [63]. These findings, consistent with our data, suggest that antimony exposure may represent a sex-specific risk factor for bone loss in women. Further research is needed to confirm these associations and elucidate the biological mechanisms underlying antimony toxicity.

Although the observed associations underscore the potential role of chronic low-level metal exposure as a modifiable risk factor for postmenopausal osteoporosis, the following limitations warrant consideration. First, the cross-sectional design precludes causal inference, and reverse causation, whereby osteoporosis influence on metal and metalloid excretion cannot be completely excluded. Second, the study population consists of postmenopausal women from a specific region of Brazil, which may limit the generalizability of the results. Third, although creatinine adjustment was applied to ensure comparability with previous studies, it may introduce bias due to variability in muscle mass [64]. This limitation should be acknowledged when interpreting the results. Finally, we did not assess dietary intake or other potential sources of metal and metalloid exposure (e.g., traffic emissions), which could also influence the observed associations.

Future longitudinal studies are essential to confirm these associations and to elucidate the underlying mechanisms. Additionally, identifying the primary sources of exposure would further support the development of effective prevention strategies.

5. Conclusions

This study adds to the growing body of evidence that environmental exposure to metals and metalloids is significantly associated with reduced BMD and increased odds of osteoporosis in postmenopausal women. Antimony demonstrated the strongest association with osteoporosis risk, while cadmium was linked to a moderate yet clinically relevant risk. While causality cannot be established, the findings highlight the need to address environmental metal exposure as part of broader strategies to preserve bone health. Future longitudinal and mechanistic studies are essential to confirm these associations and to inform risk assessment and mitigation efforts.

Supplementary Materials: The following supporting information can be downloaded at: https://www.mdpi.com/article/10.3390/toxics13060489/s1, Table S1: Comparison of participants with urinary concentrations of both Cd and Sb below and at or above the 90th percentile (P90; 1.079 and 0.729 μg/g creatinine for Cd and Sb, respectively).

Author Contributions: Conceptualization, M.C., M.C.M. and L.B.; methodology, M.C., M.C.M. and L.B.; software, M.C., M.C.M. and L.B.; validation, M.C., M.C.M. and L.B.; formal analysis, M.C.; investigation, C.T.K., B.A.R. and V.C.d.O.S.; resources, F.B.J.; data curation, M.C.; writing—original draft preparation, C.T.K.; writing—review and editing, B.A.R., F.B.J.; L.B., M.C. and M.C.M.; visualization, C.T.K. and M.C.; supervision, M.C., M.C.M. and L.B.; project administration, M.C.; funding acquisition, F.B.J. All authors have read and agreed to the published version of the manuscript.

Funding: M.C. acknowledges the support from the Portuguese National Funds (Fundação para a Ciência e Tecnologia and Ministério da Ciência, Tecnologia e Ensino Superior, FCT/MCTES) through the project UID/50006—Laboratório Associado para a Química Verde—Tecnologias e Processos Limpos. F.B.Jr thanks Fundação de Amparo à Pesquisa do Estado de São Paulo (FAPESP, 2018/24069-3 and 2021/05059-0, 2022/06443-0, 2023/03654-3, 2023/12363-2) and the Brazilian National Council for Scientific and Technological Development (CNPq 406442/2022-3) for the financial support.

Institutional Review Board Statement: The study was conducted in accordance with the Declaration of Helsinki and approved by the Research Ethics Committee of Fernando Pessoa University (Porto, Portugal), Western Paraná State University (Cascavel, Brazil), and Plataforma Brasil (approval number: 2.636.746). All individual participants provided written informed consent.

Informed Consent Statement: Informed consent was obtained from all subjects involved in the study.

Data Availability Statement: The data presented in this study are available on request from the corresponding author due to privacy restrictions.

Conflicts of Interest: The authors declare no conflicts of interest.

Abbreviations

The following abbreviations are used in this manuscript:

aOR	Adjusted odds ratio
BMI	Body mass index
DALYs	Disability-adjusted life years
DEXA	Dual-energy X-ray absorptiometry
ICP-MS	Inductively coupled plasma–mass spectrometry
HRT	Hormone replacement therapy
NHANES	U.S. National Health and Nutrition Examination Survey
WHO	World Health Organization
YLDs	Years lived with disability

References

1. Dong, Y.; Kang, H.; Peng, R.; Song, K.; Guo, Q.; Guan, H.; Zhu, M.; Ye, D.; Li, F. Global, Regional, and National Burden of Low Bone Mineral Density from 1990 to 2019: Results from the Global Burden of Disease Study 2019. *Front. Endocrinol.* **2022**, *13*, 870905. [CrossRef] [PubMed]
2. Drake, M.T.; Clarke, B.L.; Lewiecki, E.M. The Pathophysiology and Treatment of Osteoporosis. *Clin. Ther.* **2015**, *37*, 1837–1850. [CrossRef] [PubMed]
3. Compston, J.E.; McClung, M.R.; Leslie, W.D. Osteoporosis. *Lancet* **2019**, *393*, 364–376. [CrossRef]
4. Emmanuelle, N.E.; Marie-Cécile, V.; Florence, T.; Jean-François, A.; Françoise, L.; Coralie, F.; Alexia, V. Critical Role of Estrogens on Bone Homeostasis in Both Male and Female: From Physiology to Medical Implications. *Int. J. Mol. Sci.* **2021**, *22*, 1568. [CrossRef]
5. Cheng, C.H.; Chen, L.R.; Chen, K.H. Osteoporosis Due to Hormone Imbalance: An Overview of the Effects of Estrogen Deficiency and Glucocorticoid Overuse on Bone Turnover. *Int. J. Mol. Sci.* **2022**, *23*, 1376. [CrossRef]
6. Zhang, L.; Guan, Q.; Wang, Z.; Feng, J.; Zou, J.; Gao, B. Consequences of Aging on Bone. *Aging Dis.* **2023**, *15*, 2417–2452. [CrossRef] [PubMed]
7. Mi, B.; Xiong, Y.; Knoedler, S.; Alfertshofer, M.; Panayi, A.C.; Wang, H.; Lin, S.; Li, G.; Liu, G. Ageing-related bone and immunity changes: Insights into the complex interplay between the skeleton and the immune system. *Bone Res.* **2024**, *12*, 42. [CrossRef]
8. Feng, J.N.; Zhang, C.G.; Li, B.H.; Zhan, S.Y.; Wang, S.F.; Song, C.L. Global burden of hip fracture: The Global Burden of Disease Study. *Osteoporos. Int.* **2024**, *35*, 41–52. [CrossRef]
9. Foundation, I.O. Epidemiology of Osteoporosis and Fragility Fractures. Available online: https://www.osteoporosis.foundation/facts-statistics/epidemiology-of-osteoporosis-and-fragility-fractures (accessed on 11 March 2025).
10. Collaborators, G.F. Global, regional, and national burden of bone fractures in 204 countries and territories, 1990-2019: A systematic analysis from the Global Burden of Disease Study 2019. *Lancet Healthy Longev.* **2021**, *2*, e580–e592. [CrossRef]
11. Baccaro, L.F.; Conde, D.M.; Costa-Paiva, L.; Pinto-Neto, A.M. The epidemiology and management of postmenopausal osteoporosis: A viewpoint from Brazil. *Clin. Interv. Aging* **2015**, *10*, 583–591. [CrossRef]

12. Elahmer, N.R.; Wong, S.K.; Mohamed, N.; Alias, E.; Chin, K.Y.; Muhammad, N. Mechanistic Insights and Therapeutic Strategies in Osteoporosis: A Comprehensive Review. *Biomedicines* **2024**, *12*, 1635. [CrossRef] [PubMed]
13. Bjørklund, G.; Pivina, L.; Dadar, M.; Semenova, Y.; Chirumbolo, S.; Aaseth, J. Long-Term Accumulation of Metals in the Skeleton as Related to Osteoporotic Derangements. *Curr. Med. Chem.* **2020**, *27*, 6837–6848. [CrossRef] [PubMed]
14. Jalili, C.; Kazemi, M.; Taheri, E.; Mohammadi, H.; Boozari, B.; Hadi, A.; Moradi, S. Exposure to heavy metals and the risk of osteopenia or osteoporosis: A systematic review and meta-analysis. *Osteoporos. Int.* **2020**, *31*, 1671–1682. [CrossRef]
15. Tang, C.; Lv, X.; Zou, L.; Rong, Y.; Zhang, L.; Xu, M.; Li, S.; Chen, G. Cadmium exposure and osteoporosis: Epidemiological evidence and mechanisms. *Toxicol. Sci.* **2025**, *205*, 1–10. [CrossRef]
16. Briffa, J.; Sinagra, E.; Blundell, R. Heavy metal pollution in the environment and their toxicological effects on humans. *Heliyon* **2020**, *6*, e04691. [CrossRef]
17. Zhao, D.; Wang, P.; Zhao, F.J. Toxic Metals and Metalloids in Food: Current Status, Health Risks, and Mitigation Strategies. *Curr. Environ. Health Rep.* **2024**, *11*, 468–483. [CrossRef]
18. Skalny, A.V.; Aschner, M.; Silina, E.V.; Stupin, V.A.; Zaitsev, O.N.; Sotnikova, T.I.; Tazina, S.I.; Zhang, F.; Guo, X.; Tinkov, A.A. The Role of Trace Elements and Minerals in Osteoporosis: A Review of Epidemiological and Laboratory Findings. *Biomolecules* **2023**, *13*, 1006. [CrossRef] [PubMed]
19. Wang, C.; Zhu, Y.; Long, H.; Ou, M.; Zhao, S. Relationship between blood manganese and bone mineral density and bone mineral content in adults: A population-based cross-sectional study. *PLoS ONE* **2022**, *17*, e0276551. [CrossRef]
20. Molenda, M.; Kolmas, J. The Role of Zinc in Bone Tissue Health and Regeneration-a Review. *Biol. Trace Elem. Res.* **2023**, *201*, 5640–5651. [CrossRef]
21. Rayman, M.P. The importance of selenium to human health. *Lancet* **2000**, *356*, 233–241. [CrossRef]
22. Yang, T.; Lee, S.Y.; Park, K.C.; Park, S.H.; Chung, J.; Lee, S. The Effects of Selenium on Bone Health: From Element to Therapeutics. *Molecules* **2022**, *27*, 392. [CrossRef] [PubMed]
23. Razzaque, M.S.; Wimalawansa, S.J. Minerals and Human Health: From Deficiency to Toxicity. *Nutrients* **2025**, *17*, 454. [CrossRef]
24. Wei, M.H.; Cui, Y.; Zhou, H.L.; Song, W.J.; Di, D.S.; Zhang, R.Y.; Huang, Q.; Liu, J.A.; Wang, Q. Associations of multiple metals with bone mineral density: A population-based study in US adults. *Chemosphere* **2021**, *282*, 131150. [CrossRef] [PubMed]
25. Engström, A.; Michaëlsson, K.; Suwazono, Y.; Wolk, A.; Vahter, M.; Akesson, A. Long-term cadmium exposure and the association with bone mineral density and fractures in a population-based study among women. *J. Bone Miner. Res.* **2011**, *26*, 486–495. [CrossRef]
26. Li, T.; Xie, Y.; Wang, L.; Huang, G.; Cheng, Y.; Hou, D.; Liu, W.; Zhang, T.; Liu, J. The Association between Lead Exposure and Bone Mineral Density in Childhood and Adolescence: Results from NHANES 1999–2006 and 2011–2018. *Nutrients* **2022**, *14*, 1523. [CrossRef]
27. Kunioka, C.T.; Manso, M.C.; Carvalho, M. Association between Environmental Cadmium Exposure and Osteoporosis Risk in Postmenopausal Women: A Systematic Review and Meta-Analysis. *Int. J. Environ. Res. Public Health* **2022**, *20*, 485. [CrossRef]
28. Wang, W.J.; Wu, C.C.; Jung, W.T.; Lin, C.Y. The associations among lead exposure, bone mineral density, and FRAX score: NHANES, 2013 to 2014. *Bone* **2019**, *128*, 115045. [CrossRef]
29. Yachiguchi, K.; Sekiguchi, T.; Nakano, M.; Hattori, A.; Yamamoto, M.; Kitamura, K.; Maeda, M.; Tabuchi, Y.; Kondo, T.; Kamauchi, H.; et al. Effects of inorganic mercury and methylmercury on osteoclasts and osteoblasts in the scales of the marine teleost as a model system of bone. *Zoolog. Sci.* **2014**, *31*, 330–337. [CrossRef] [PubMed]
30. Cho, G.J.; Park, H.T.; Shin, J.H.; Hur, J.Y.; Kim, S.H.; Lee, K.W.; Kim, T. The relationship between blood mercury level and osteoporosis in postmenopausal women. *Menopause* **2012**, *19*, 576–581. [CrossRef]
31. Mizuno, D.; Kawahara, M.; Konoha-Mizuno, K.; Ogawara, T.; Hama, R.; Yamazaki, K. Toxic Effects of Two Redox States of Thallium on Immortalised Hypothalamic GT1-7 Neuronal Cells. *Int. J. Mol. Sci.* **2023**, *24*, 1583. [CrossRef]
32. Korotkov, S.M. Mitochondrial Oxidative Stress Is the General Reason for Apoptosis Induced by Different-Valence Heavy Metals in Cells and Mitochondria. *Int. J. Mol. Sci.* **2023**, *24*, 4459. [CrossRef] [PubMed]
33. Gebel, T. Arsenic and antimony: Comparative approach on mechanistic toxicology. *Chem. Biol. Interact.* **1997**, *107*, 131–144. [CrossRef] [PubMed]
34. Zamani, A.; Omrani, G.R.; Nasab, M.M. Lithium's effect on bone mineral density. *Bone* **2009**, *44*, 331–334. [CrossRef]
35. Liu, B.; Wu, Q.; Zhang, S.; Del Rosario, A. Lithium use and risk of fracture: A systematic review and meta-analysis of observational studies. *Osteoporos. Int.* **2019**, *30*, 257–266. [CrossRef] [PubMed]
36. Wong, S.K.; Chin, K.Y.; Ima-Nirwana, S. The Skeletal-Protecting Action and Mechanisms of Action for Mood-Stabilizing Drug Lithium Chloride: Current Evidence and Future Potential Research Areas. *Front. Pharmacol.* **2020**, *11*, 430. [CrossRef]
37. Wang, X.; Qu, Z.; Zhao, S.; Luo, L.; Yan, L. Wnt/β-catenin signaling pathway: Proteins' roles in osteoporosis and cancer diseases and the regulatory effects of natural compounds on osteoporosis. *Mol. Med.* **2024**, *30*, 193. [CrossRef]
38. Kołodziejska, B.; Stępień, N.; Kolmas, J. The Influence of Strontium on Bone Tissue Metabolism and Its Application in Osteoporosis Treatment. *Int. J. Mol. Sci.* **2021**, *22*, 6564. [CrossRef]

39. Marx, D.; Rahimnejad Yazdi, A.; Papini, M.; Towler, M. A review of the latest insights into the mechanism of action of strontium in bone. *Bone Rep.* **2020**, *12*, 100273. [CrossRef]
40. Gritsaenko, T.; Pierrefite-Carle, V.; Creff, G.; Simoneau, B.; Hagège, A.; Farlay, D.; Pagnotta, S.; Orange, F.; Jaurand, X.; Auwer, C.D.; et al. Low doses of uranium and osteoclastic bone resorption: Key reciprocal effects evidenced using new in vitro biomimetic models of bone matrix. *Arch. Toxicol.* **2021**, *95*, 1023–1037. [CrossRef]
41. Arzuaga, X.; Gehlhaus, M.; Strong, J. Modes of action associated with uranium induced adverse effects in bone function and development. *Toxicol. Lett.* **2015**, *236*, 123–130. [CrossRef]
42. Tissandie, E.; Guéguen, Y.; Lobaccaro, J.M.; Grandcolas, L.; Grison, S.; Aigueperse, J.; Souidi, M. Vitamin D metabolism impairment in the rat's offspring following maternal exposure to 137cesium. *Arch. Toxicol.* **2009**, *83*, 357–362. [CrossRef] [PubMed]
43. Barbosa, F., Jr.; Devoz, P.P.; Cavalcante, M.R.N.; Gallimberti, M.; Cruz, J.C.; Domingo, J.L.; Simões, E.J.; Lotufo, P.; Liu, S.; Bensenor, I. Urinary levels of 30 metal/metalloids in the Brazilian southeast population: Findings from the Brazilian Longitudinal Study of Adult Health (ELSA-Brasil). *Environ. Res.* **2023**, *225*, 115624. [CrossRef]
44. Nishad, P.A.; Bhaskarapillai, A. Antimony, a pollutant of emerging concern: A review on industrial sources and remediation technologies. *Chemosphere* **2021**, *277*, 130252. [CrossRef]
45. Esteban, M.; Castaño, A. Non-invasive matrices in human biomonitoring: A review. *Environ. Int.* **2009**, *35*, 438–449. [CrossRef]
46. Shilnikova, N.; Momoli, F.; Karyakina, N.; Krewski, D. Review of non-invasive biomarkers as a tool for exposure characterization in human health risk assessments. *J. Toxicol. Environ. Health B Crit. Rev.* **2025**, *28*, 122–150. [CrossRef]
47. Gil, F.; Hernández, A.F. Toxicological importance of human biomonitoring of metallic and metalloid elements in different biological samples. *Food Chem. Toxicol.* **2015**, *80*, 287–297. [CrossRef] [PubMed]
48. Martinez-Morata, I.; Sobel, M.; Tellez-Plaza, M.; Navas-Acien, A.; Howe, C.G.; Sanchez, T.R. A State-of-the-Science Review on Metal Biomarkers. *Curr. Environ. Health Rep.* **2023**, *10*, 215–249. [CrossRef] [PubMed]
49. Herzberg, M.; Foldes, J.; Steinberg, R.; Menczel, J. Zinc excretion in osteoporotic women. *J. Bone Miner. Res.* **1990**, *5*, 251–257. [CrossRef]
50. Kunioka, C.T.; Cruz, J.C.; Souza, V.C.O.; Rocha, B.A.; Barbosa Jr, F.; Belo, L.; Manso, M.C.; Carvalho, M. Low-Level Environmental Cadmium Exposure and Its Effects on Renal and Bone Health in Brazilian Postmenopausal Women: A Cross-Sectional Study. *Expo. Health* **2025**. [CrossRef]
51. Qin, L.; Liu, Q.; Zhang, T.; Tang, X.; Mo, X.; Liang, Y.; Wang, X.; Cao, J.; Huang, C.; Lu, Y.; et al. Association Between Combined Polymetallic Exposure and Osteoporosis. *Biol. Trace Elem. Res.* **2024**, *202*, 3945–3958. [CrossRef]
52. Haider, F.U.; Zulfiqar, U.; Ain, N.U.; Mehmood, T.; Ali, U.; Ramos Aguila, L.C.; Li, Y.; Siddique, K.H.M.; Farooq, M. Managing antimony pollution: Insights into Soil-Plant system dynamics and remediation Strategies. *Chemosphere* **2024**, *362*, 142694. [CrossRef]
53. Zhao, S.; Shi, T.; Terada, A.; Riya, S. Evaluation of Pollution Level, Spatial Distribution, and Ecological Effects of Antimony in Soils of Mining Areas: A Review. *Int. J. Environ. Res. Public Health* **2022**, *20*, 242. [CrossRef] [PubMed]
54. Kim, E.S.; Shin, S.; Lee, Y.J.; Ha, I.H. Association between blood cadmium levels and the risk of osteopenia and osteoporosis in Korean post-menopausal women. *Arch. Osteoporos.* **2021**, *16*, 22. [CrossRef]
55. Wang, M.; Wang, X.; Liu, J.; Wang, Z.; Jin, T.; Zhu, G.; Chen, X. The Association Between Cadmium Exposure and Osteoporosis: A Longitudinal Study and Predictive Model in a Chinese Female Population. *Front. Public Health* **2021**, *9*, 762475. [CrossRef] [PubMed]
56. Wu, L.; Wei, Q.; Lv, Y.; Xue, J.; Zhang, B.; Sun, Q.; Xiao, T.; Huang, R.; Wang, P.; Dai, X.; et al. Wnt/β-Catenin Pathway Is Involved in Cadmium-Induced Inhibition of Osteoblast Differentiation of Bone Marrow Mesenchymal Stem Cells. *Int. J. Mol. Sci.* **2019**, *20*, 1519. [CrossRef]
57. Luo, H.; Gu, R.; Ouyang, H.; Wang, L.; Shi, S.; Ji, Y.; Bao, B.; Liao, G.; Xu, B. Cadmium exposure induces osteoporosis through cellular senescence, associated with activation of NF-κB pathway and mitochondrial dysfunction. *Environ. Pollut.* **2021**, *290*, 118043. [CrossRef] [PubMed]
58. Youness, E.R.; Mohammed, N.A.; Morsy, F.A. Cadmium impact and osteoporosis: Mechanism of action. *Toxicol. Mech. Methods* **2012**, *22*, 560–567. [CrossRef]
59. Wang, J.; Xue, M.; Hu, Y.; Li, J.; Li, Z.; Wang, Y. Proteomic Insights into Osteoporosis: Unraveling Diagnostic Markers of and Therapeutic Targets for the Metabolic Bone Disease. *Biomolecules* **2024**, *14*, 554. [CrossRef]
60. Galvez-Fernandez, M.; Rodriguez-Hernandez, Z.; Grau-Perez, M.; Chaves, F.J.; Garcia-Garcia, A.B.; Amigo, N.; Monleon, D.; Garcia-Barrera, T.; Gomez-Ariza, J.L.; Briongos-Figuero, L.S.; et al. Metabolomic patterns, redox-related genes and metals, and bone fragility endpoints in the Hortega Study. *Free Radic. Biol. Med.* **2023**, *194*, 52–61. [CrossRef]
61. Ximenez, J.P.B.; Zamarioli, A.; Kacena, M.A.; Barbosa, R.M.; Barbosa, F., Jr. Association of Urinary and Blood Concentrations of Heavy Metals with Measures of Bone Mineral Density Loss: A Data Mining Approach with the Results from the National Health and Nutrition Examination Survey. *Biol. Trace Elem. Res.* **2021**, *199*, 92–101. [CrossRef]

62. Rana, J.; Renzetti, S.; Islam, R.; Donev, M.; Hu, K.; Oulhote, Y. Mediation Effect of Metal Mixtures in the Association Between Socioeconomic Status and Self-rated Health Among US Adults: A Weighted Quantile Sum Mediation Approach. *Expo. Health* **2022**, *14*, 609–621. [CrossRef]
63. Li, H.; Li, G.; Yi, M.; Zhou, J.; Deng, Y.; Huang, Y.; He, S.; Meng, X.; Liu, L. Sex-specific associations of urinary mixed-metal concentrations with femoral bone mineral density among older people: An NHANES (2017–2020) analysis. *Front. Public Health* **2024**, *12*, 1363362. [CrossRef] [PubMed]
64. Suwazono, Y.; Akesson, A.; Alfvén, T.; Järup, L.; Vahter, M. Creatinine versus specific gravity-adjusted urinary cadmium concentrations. *Biomarkers* **2005**, *10*, 117–126. [CrossRef] [PubMed]

toxics

MDPI

Article

Integrated Deterministic and Probabilistic Methods Reveal Heavy Metal-Induced Health Risks in Guizhou, China

Qinju Li †, Dashuan Li †, Zelan Wang, Dali Sun, Ting Zhang * and Qinghai Zhang *

School of Public Health, Guizhou Medical University, Guiyang 561113, China; leeqinju@163.com (Q.L.); lidashuan1997@163.com (D.L.); wzl2010kaoyan@163.com (Z.W.); dalisun11@163.com (D.S.)
* Correspondence: zhangt_17@163.com (T.Z.); zhqh8@163.com (Q.Z.)
† These authors contributed equally to this work.

Abstract: Due to high geological background and intensive mining activities, soils are prone to heavy metals (HMs) accumulation and ecological fragility in Guizhou Province, China. A total of 740 topsoil samples were therefore collected, and aimed to determine the concentrations of As, Cd, Cr, Hg, and Pb, estimate the ecological pollution, and evaluate the carcinogenic and non-carcinogenic health risks to humans. Results showed As (1.08%) and Cd (24.46%) in soil exceeded standards. The Igeo showed that Cr (1.49%) and Hg (31.62%) in soil were at light pollution levels; single factor pollution index (PI) showed that Cd (21.35%) in soil was mildly polluted; risk index (RI) as at a low risk level. Notably, both deterministic and Monte Carlo analyses revealed unacceptable carcinogenic risks for As and Cr in children, with traditional methods potentially underestimating As risks. Moreover, Target-Organ Toxicity Dose (TTD) revealed soil HMs as a higher risk to hematological health, with notable health risks posed by Pb in children. It is noted that spatial distribution analysis suggested that the southwestern region of Guizhou Province should be prioritized for health risk management and control. By integrating the uniqueness of geological environments, multi-dimensional health risk assessments, and spatial distributions, the present study provides a scientific basis for assessing HMs pollution risks and soil health risks in the karst regions.

Keywords: heavy metals; agricultural soil; health risk assessments; target organ toxic dose method

1. Introduction

Heavy metals (HMs) contamination in soil emerged as a critical environmental problem due to their inherent toxicity, persistent retention, non-degradability, and sustained bioavailability [1,2]. Soil is an important carrier of HMs. Excessive HMs can disrupt the productivity and quality of soil when entering into the soil layer [3]. Furthermore, severely contaminated soils may serve as a persistent source of groundwater and ecosystem contamination [4]. More seriously, when the concentration of HMs such as lead (Pb), arsenic (As), and mercury (Hg) in the farmland soil reaches a certain level, it will destroy the internal balance of the farmland [5]. It is important to note that soil ecosystems, particularly those of farmland, are crucial for human survival and development [6]. It is essential to evaluate the contamination status and ecological risks associated with HMs contamination in agroecosystems.

Soil contamination by HMs varies across China, with southern provinces bearing the brunt of the problem; these regions should be considered key areas for monitoring and managing HM pollution. Simultaneously, cadmium (Cd), Hg, Pb, chromium (Cr), and arsenic (As) have been identified as the main HMs requiring targeted control efforts within Chinese soils [7]. Therefore, this study focuses on these five HMs. In, addition, Southwest China constitutes a pivotal component of karst region all over the world [8]. Karst area is a fragile ecosystem prone to human impacts with high geochemical background value and limited HM capacity. Karst landscapes, formed from carbonate rock formations, are among the regions naturally characterized by elevated HM concentrations due to their unique geological composition. Research has shown that as these carbonate-rich areas undergo weathering, essential elements like calcium and magnesium leach out, leading to either the preservation or further accumulation of HMs. HMs typically accumulate in the residue, showing a steady proportional rise in both concentration and volume. This persistent buildup allows them to remain detectable in regions with minimal bedrock composition, even following extensive soil formation processes [9,10]. Geological surveys reveal that carbonate rock formations in Guizhou Province cover approximately 1.1×10^5 km^2, which accounts for 73% of the provincial territory [11] and ranks first in China. And, the researches reported that mining and smelting in Southwest China could release HMs to cause superimposed soil pollution [12]. Presently, these techniques of single factor pollution index (PI), Nemero pollution index (P_N), geo-accumulation index (Igeo), ecological risk (Er) and risk index (RI) were all conducted to investigate the ecological risks polluted by HMs in karst areas [13,14]. Tang et al. [15] found that P_N was moderately polluted for the agricultural soils of karst areas. Qin et al. [16] indicated that Er in the karst area of Yunnan Province reached moderate risk accounting for 55.27% of the total samples. Therefore, the ecological pollution status of HMs in the karst areas deserves further attention.

The health risk assessment, a quantitative approach, evaluates potential human exposure risks through ingestion, dermal contact, and inhalation [17,18]. Specifically, the concentration of HMs and exposure parameters were the main considerations for deterministic risk assessment [19,20]. However, the deterministic risk, calculated by fixed values, relies on the actual magnitude of the risk defined by individual differences, age, physical condition, gender, and metabolic parameters [21,22]. In contrast, probabilistic risk simulation provided a more accurate basis for risk management and remediation [23]. Based on probabilistic modeling, the Monte Carlo simulation technique incorporates the variability in critical exposure traits, such as the fluctuation in soil ingestion rate (R_{ing}), body weight (BW) and exposure frequency (EF). In probabilistic risk assessment, every variable and parameter is treated as a probability distribution rather than a fixed value. This approach dramatically minimizes the uncertainties inherent in health risk evaluations. The Monte Carlo simulation method follows these essential steps: First, probability distributions are established for HM concentration parameters. Next, uncertain parameter values are generated according to these distributions. The model then performs 10,000 randomized samplings across these parameter ranges, feeding them into the risk assessment calculations. Finally, the simulated output parameters are analyzed to generate cumulative probability distributions that quantify potential health risks [24,25]. Therefore, Monte Carlo simulation can well make up for the deficiencies of classical techniques. This technique can estimate the probability of pollutants exceeding the danger threshold and prioritize the part of health risk control, effectively conducting probabilistic health risk analysis [26]. For example, Eslami et al. [27] studied the health risks of pesticides on fruits, and the Monte Carlo simulation they adopted revealed that the total hazard quotient (THQ = 36.7%) of children was significantly higher than that of adults (7.8%). Traditional mean analysis was unable to capture this difference, demonstrating the precise identification ability of probability methods for sensitive populations. The emphasis suggested that evaluating probabilistic risk could yield a more suitable health risk assessment, to some degree [28]. In addition, the Target-Organ Toxicity Dose (TTD), a method for risk characterization of specific toxicological endpoints, which is an improvement on the traditional health risk assessment. It not only accounts for the critical effects of pollutants but also integrates the assessment of toxic doses across multiple target organs for diverse HMs, thereby significantly enhancing the precision of risk evaluations. To some extent, it places a particular emphasis on the potential impacts on target organs when pollutant concentrations surpass critical exposure doses [29]. Still, it is limited to establishing a more comprehensive and multi-perspective assessment on integrating deterministic risk, probabilistic risk and target organ toxicity dose to pay attention to human health.

As the discussion above, our study aimed to (1) systematically evaluate the spatial distribution and contamination severity of HMs; (2) assess the carcinogenic and non-carcinogenic health risks of HMs using a deterministic assessment and the Monte Carlo method; and (3) estimate the non-carcinogenic health risks of HMs using the TTD method. The present study can provide a theoretical framework for the scientific evaluation and human health risk assessment of HMs contamination in karst areas of Guizhou Province.

2. Materials and Methods

2.1. Study Area

Guizhou Province (24°37′ N–29°13′ N, 103°36′ E–109°35′ E), lies within the eastern segment of the Yunnan-Guizhou Plateau, featuring elevated ground to the west and descending terrain to the east. The annual temperatures hover between 14–16 °C and rainfall typically ranges from 1100–1400 mm. Spanning 176,167 square kilometers, this vital agricultural hub supports a population of around 35.81 million people and plays a key role in China's grain production [30].

2.2. Sample Collection and Analysis

In autumn 2017, 740 topsoil samples (from 0–20 cm depth) were gathered from Guizhou's farmland, nearly encompassing the entire cultivated area. The layout of sampling points referred to the Chinese national standard DZ/T 0295-2016 [31], soil samples were collected in Guiyang (GY, n = 48), Zunyi (ZY, n = 148), Anshun (AS, n = 82), Liupanshui (LPS, n = 21), Bijie (BJ, n = 51), Qiandongnan (QDN, n = 151), Qiannan (QN, n = 119), Qianxinan (QXN, n = 44) and Tongren (TR, n = 75). The distributions of sampling sites were shown in Figure 1.

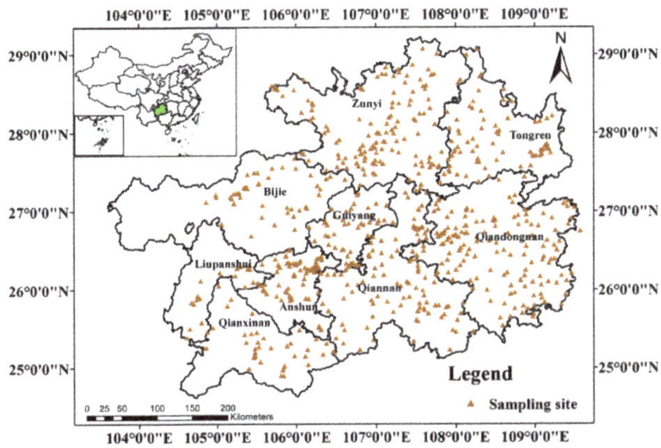

Figure 1. Location of the study area and distribution of sampling sites.

To ensure a representative composite sample, five individual subsamples were gathered within a roughly 10-m radius of the target site and carefully homogenized. Following collection, all specimens were left to air-dry under ambient laboratory conditions. The soil sample processing procedure was conducted as follows: First, a 2 mm sieve was employed to remove animal residues, stones, and plant materials. The ground soil sample was then reduced to approximately 400 g using the quartering method. Subsequently, the entire sample was uniformly sieved through a 0.25 mm sieve. Of the resulting material, one-quarter was allocated for soil pH measurement, while the remaining portion (300 g) was further sieved through a 0.15 mm sieve for the determination of HMs in the soil [32]. Subsequently, 0.25 g of soil was transferred into a Teflon crucible, and then a 10 mL mixture of nitric acid and perchloric acid in a 4:1 ratio, along with 2 mL of hydrofluoric acid, was added. The crucible was subsequently heated to promote digestion. Once digested, the solution was moved to a 25 mL colorimetric tube and topped up with ultra-pure water to the required volume. The Pb, Cd, and Cr contents in the soil were then measured using Inductively Coupled Plasma-Mass Spectrometry (ICP-MS) [33]. ICP-MS allows for simultaneous multi-element detection, offering rapid, highly sensitive analysis of trace amounts [34]. In another process, about 0.3 g of soil sample via a 0.15 mm sieve was weighed and placed into a 50 mL colorimetric tube. The sample was slightly wetted with water and then mixed with 10 mL of a 1:1 aqua regia solution (HCl-HNO_3-H_2O, 3:1:4). This mixture was digested in a boiling water bath for 2 h, then allowing to cool, and diluting with ultra-pure water to the desired volume. The Hg content was measured using an Atomic Fluorescence Spectrometer (AF-630A) [35]. For As determination, the aliquot was taken and mixed with thiourea and ascorbic acid [36]. Hg and As were detected by AFS, and this method has high sensitivity and accuracy [37].

To measure the pH of soils, about 10 g of soil were carefully weighed and transferred into a 50 mL beaker. Then, 25 mL of water was poured in to eliminate any trapped carbon dioxide. The mixture was stirred vigorously to ensure thorough blending and left to sit for half an hour. Finally, the pH level was measured using a glass electrode.

During the determination process, GBW07408 (from the National Standards Research Center of China) was utilized for HM content analysis and quality control. The recovery rate of spiked samples was maintained between 90% and 110%. Parallel samples were taken every 30 samples, with relative deviations kept within 10%.

2.3. Contamination Assessment

The PI, P_N and Igeo were applied to assess the contamination of HMs in farmland in Guizhou Province. The PI assessed individual HM pollutant concentrations in the soils [38], while the P_N gauged the overall pollution impact of multiple HMs [39]. These indices are calculated as follows:

$$PI = \frac{C_i}{S_i} \tag{1}$$

$$P_N = \sqrt{\frac{PI_{ave}^2 + PI_{max}^2}{2}} \tag{2}$$

where C_i is the measured concentration of HM i, S_i is the standard evaluation value, which is the soil pollution risk screening value in the "Soil environmental quality standard" [40]. The pollution level classifications for PI and P_N were shown in Table S1 [41].

Considering the impact of natural diagenesis on the background values, the geological accumulation index (Igeo) was employed to assess the contamination of HMs, and it identified the influence of anthropogenic activities [42]. It was calculated as follows:

$$I_{geo} = \log_2\left[\frac{C_i}{(1.5 \times B_i)}\right] \tag{3}$$

where B_i is the geochemical background value of HM i in the local soil [43], C_i is the measured concentration of HM i, and 1.5 is the coefficient of variation that results from rock formation [44]. The pollution level classification for Igeo was shown in Table S2 [42].

2.4. Potential Ecological Risk Assessment

The potential ecological risk index method proposed by Hakanson [38] can comprehensively consider the ecotoxicity of pollutants and ecological environmental factors. This approach effectively captures the overall influence of different contaminants on the ecosystem. This index can be used to quantitatively analyze and predict potential ecological risks. RI is the sum of the ecological health risk index for each HM (Er), calculated as follows:

$$E_r = T_i \times \left(\frac{C_i}{S_i}\right) \tag{4}$$

$$RI = \sum E_r \tag{5}$$

where T_i is the toxic response factor (As: 10, Cd: 30, Cr: 2, Hg: 40, Pb: 5). The risk levels classifications for Er and RI were shown in Table S3 [38].

2.5. Health Risk Assessment

The health risk assessment recommended by the USEPA for human exposure to HM was quantified both non-carcinogenic and carcinogenic risks via oral ingestion (ing), inhalation (inh) and dermal contact (dermal), respectively [45]:

$$ADD_{ing} = C_S \times \frac{R_{ing} \times EF \times ED}{BW \times AT} \times 10^{-6} \tag{6}$$

$$ADD_{inh} = C_S \times \frac{R_{inh} \times EF \times ED}{PEF \times BW \times AT} \tag{7}$$

$$ADD_{der} = C_S \times \frac{AF \times SA \times ABS \times EF \times ED}{BW \times AT} \times 10^{-6} \tag{8}$$

where ADD_{ing}, ADD_{inh}, and ADD_{der} represent the average daily doses of HMs in the soil in mg/(kg·d); C_S is the soil HM concentration (mg/kg). The interpretation and values of the exposure parameters are shown in Table S4 [46–48].

The formulas below were used to compute the non-carcinogenic and carcinogenic risk indices:

$$HI_i = \sum \frac{ADD_i}{RfD_i} \tag{9}$$

$$THI = \sum HI_i = \sum (HI_{ing} + HI_{inh} + HI_{der}) \tag{10}$$

$$CR_i = \sum (ADD_i \times SF) \tag{11}$$

$$TCR = \sum CR_i \tag{12}$$

HI represents the non-cancerous health risk associated with a single HM across various exposure routes, while the THI aggregates the risks from multiple HMs. If either the THI or HI exceeds 1, it signals a possible risk of health problems [49]. RfD_i indicates the non-carcinogenic average daily reference dose for HM i. CR is the carcinogenic risk factor for all exposure pathways for a single HM, the TCR indicates the total carcinogenic risk for multiple HMs. SF is a carcinogenic slope factor. When the CR values surpass the risk cutoff of 1×10^{-4}, it suggests that humans face significant risks of cancer. Conversely, if the CR values fall below the commonly accepted threshold of 1×10^{-6}, they are typically viewed as posing an insignificant threat to human health [50]. The values of exposure parameters relevant to adults and children in the health risk assessment are shown in Table S5 [46–48,51].

To address uncertainties and variability in risk quantification, a probabilistic framework employing Monte Carlo simulations was implemented. Computational analyses were conducted using Oracle Crystal Ball, with 10,000 iterative samplings at a 95% confidence interval, drawing stochastically from predefined exposure parameter distributions [52]. This approach generated probabilistic health risk profiles, while the parameter configurations for probability density functions in the risk assessment model [53] were detailed in Table S6 [46,54–56].

In addition, TTD is an improvement on the HI method, and the toxic dose of HM in multiple target organs is included in the evaluation scope, which can more accurately reflect the specific health risks of pollutants to humans [57]. At present, the corresponding target organ toxicity data for Cd, Pb, As and Cr are relatively complete, and the corresponding target organ toxicity effect endpoint data have been reported, while the target organ toxicity data for Hg are relatively lacking [58]. The formula for calculating HI in the TTD method is as follows:

$$HI_{TTD} = \sum \frac{ADD_i}{TTD_i} \quad (13)$$

$$THI_{TTD} = \sum HI_i \quad (14)$$

where the TTD_i value is the endpoint of the toxicological effect of the corresponding target organ for each HM (Table S7) [59]. HI_{TTD} is the risk value of a single HM to the target organ. The THI_{TTD} is the sum of the HI_{TTD} of multiple HMs.

2.6. Data Analysis

Microsoft Office Excel 2024, IBM SPSS Statistics 27, Origin 2024 and GraphPad Prism 10 were employed for experimental data processing and analysis. Independent sample *t*-test of health risk indices (HI, CR) between adults and children were performed using Student's *t*-test. Subsequently, Kriging interpolation of ArcGIS 10.8 software was used to conduct spatial interpolation mapping to describe the spatial distribution of HMs.

3. Results

3.1. Evaluation of Heavy Metal Pollution

The results of HMs pollution in soil were shown in Table 1. The mean values of As, Cd, Cr, Hg and Pb for the soil samples were 9.08, 0.36, 73.06,0.13 and 28.14 mg/kg, respectively. When compared with the soil background values of Guizhou Province, the exceeding rates of As, Cd, Cr, Hg and Pb were 4.46%, 5.54%, 20.95%, 62.03% and 18.92%, respectively. Furthermore, the soils in Guizhou Province mainly exhibited slight acidity, with pH values between 3.84 and 8.06 and an average of 6.14. At pH levels ≤ 5.5, between 5.5 and 6.5, and from 6.5 to 7.5, Cd exceedance rates were 37.39%, 32.03%, and 9.47%, respectively. At the pH > 7.5, the exceeding rate of As was 9.88%. The total exceedance rates of As and Cd were 1.08% and 24.46%, respectively. Moreover, the coefficient of variation (CV) was in the order of As (59.25%) > Hg (46.15%) > Cd (44.44%) > Cr (39.98%) > Pb (28.75%).

The kriging method was further used to analyze the current status and spatial distribution of HM contamination in agricultural soils (Figure 2). High levels of As and Hg were observed in the central area; high Cd and Cr were concentrated mainly in the western region; and Pb concentrations were higher from the northern parts, respectively. And the levels of pH were high in the southwestern part.

Table 1. Concentrations of heavy metal in soils in the study area (mg/kg).

Items	As	Cd	Cr	Hg	Pb	pH
Min	1.19	0.08	15.97	0.02	11.42	3.84
Max	24.87	0.75	149.82	0.29	49.38	8.06
Mean	9.08	0.36	73.06	0.13	28.14	6.14
SD	5.38	0.16	29.21	0.06	8.09	0.96
CV%	59.25	44.44	39.98	46.15	28.75	15.64
pH ≤ 5.5	30	0.30	250	0.50	80	-
Exceeded(%) [a]	0	37.39	0	0	0	-
5.5 < pH ≤ 6.5	30	0.4	250	0.50	100	-
Exceeded(%) [a]	0	32.03	0	0	0	-
6.5 < pH ≤ 7.5	25	0.6	300	0.6	140	-
Exceeded(%) [a]	0	9.47	0	0	0	-
pH > 7.5	20	0.8	350	1.0	240	-
Exceeded(%) [a]	9.88	0	0	0	0	-
BV [b]	20	0.66	95.9	0.1	35.2	-
Exceeded(%) [b]	4.46	5.54	20.95	62.03	18.92	-

[a] Soil pollution risk screening values refer to "Soil environmental quality standard" (GB15618-2018) [40]. [b] China National Environmental Monitoring Center (CNEMC), the Backgrounds of Soil Environment of Guizhou, China.

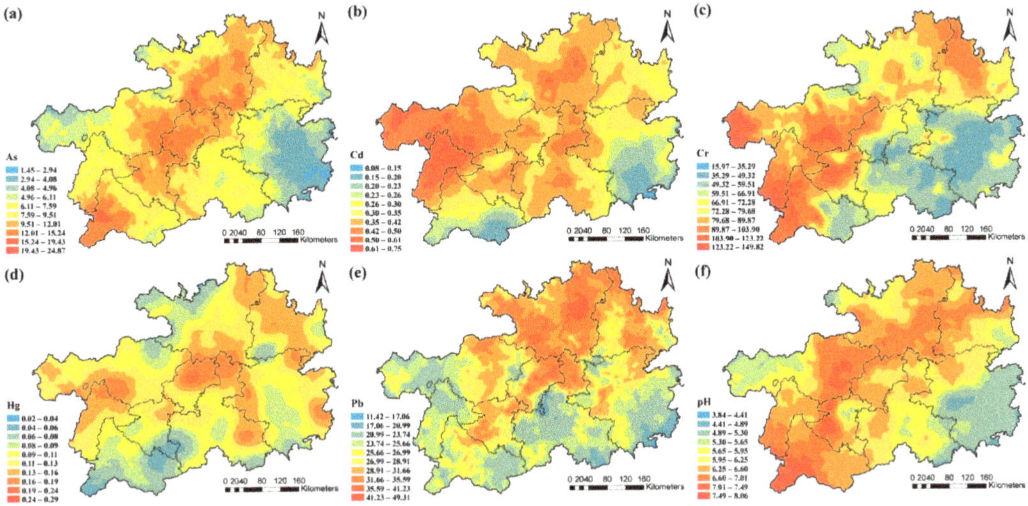

Figure 2. Spatial distribution of heavy metal pollution characteristics. (**a**) As; (**b**) Cd; (**c**) Cr; (**d**) Hg; (**e**) Pb; (**f**) pH.

In addition, the methods of Igeo and PI were utilized to assess the contamination status of farmland soils. assess the contamination status of farmland soils. in the Guizhou Province. The values of Igeo indicated that 1.49% of Cr and 31.62% of Hg, indicating minor pollution (Figure 3a). The average PI for HMs varied between 0.23 and 0.84. Specifically, 21.35% of the sampling sites exhibited mild Cd contamination (1 < PI ≤ 2), and 3.11% showed moderate Cd contamination (2 < PI ≤ 3) (Figure 3b). The P_N value was 1.77, which was at the light pollution level.

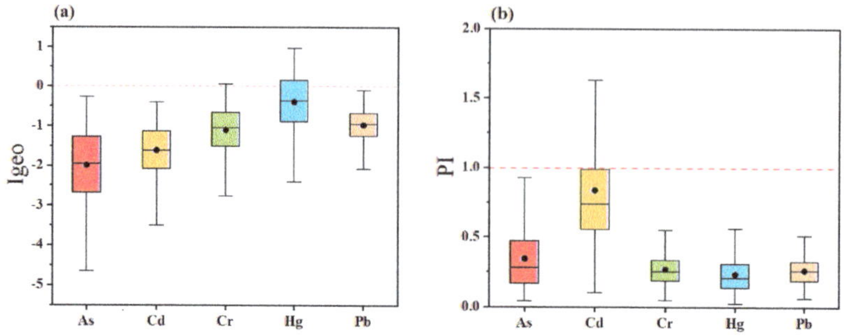

Figure 3. Assessment of the (**a**) Igeo and (**b**) PI.

3.2. Ecological Risk Assessment

The Er values calculated by HMs concentrations rank as the order of Cd > Pb > As > Hg > Cr, with 10.95% sites of Cd rated at a moderate risk level (Figure 4a). And, there was no ecological risk level in Guizhou because the RI value was below 80. More importantly, the spatial pattern of RI (Figure 4b) indicated higher concentrations in the western area, with slightly higher values in the central region.

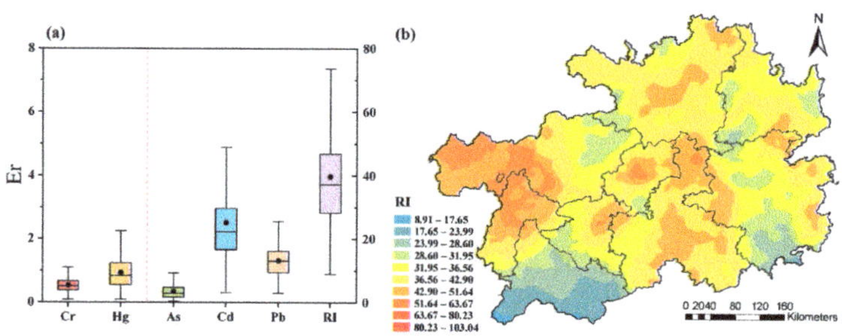

Figure 4. Assessment of (**a**) Er and spatial distribution of (**b**) RI.

3.3. Deterministic Risk Assessment

The results of the deterministic risk assessment were shown in Table 2. The results for the three exposure pathways were in the order of $HI_{ing}(1.12 \times 10^{-1}) > HI_{dermal}(1.54 \times 10^{-2}) > HI_{inh}(7.62 \times 10^{-4})$ in adults and $HI_{ing}(7.54 \times 10^{-1}) > HI_{dermal}(7.22 \times 10^{-2}) > HI_{inh}(1.32 \times 10^{-3})$ in children, indicating that oral ingestion is the primary exposure pathway for non-carcinogenic risks. The deterministic risks posed by HMs, the order of HI was $Cr(5.12 \times 10^{-2}) > As(4.59 \times 10^{-2}) > Pb(3.03 \times 10^{-2}) > Hg(6.82 \times 10^{-4}) > Cd(6.42 \times 10^{-4})$ in adults and $Cr(3.12 \times 10^{-1}) > As(3.05 \times 10^{-1}) > Pb(2.02 \times 10^{-1}) > Hg(4.50 \times 10^{-3}) > Cd(4.03 \times 10^{-3})$ in children. Although all soil HI values were under 1, 27.84% THI values in children and 30% TCR values in adults were above the acceptable range. At the same time, the order for CR was $Cr(6.32 \times 10^{-5}) > As(2.04 \times 10^{-5}) > Cd(3.27 \times 10^{-6}) > Pb(3.56 \times 10^{-7})$ in adults and $Cr(4.06 \times 10^{-4}) > As(1.36 \times 10^{-4}) > Cd(2.19 \times 10^{-5}) > Pb(2.39 \times 10^{-6})$ in children. CR(Cr) exceeded the acceptable range in 9.59% of adults. Specifically, the CRing values of As and Cr for children in the ingestion pathway were greater than 1×10^{-4}. And, the exceeding rate of CR(As) in children was 57.30%. The Student's t-test showed that there were significant differences in all health risk indicators between children and adults (Figures S1 and S2). For THI, TCR, As, Cd, Cr, Hg and Pb, the non-carcinogenic and carcinogenic risks were notably greater in children compared to adults ($p < 0.0001$).

Table 2. Carcinogenic and non-carcinogenic risk assessment of heavy metals in different populations.

		Non-Carcinogenic Risks				Carcinogenic Risks			
		HI_{ing}	HI_{inh}	HI_{dermal}	HI	CR_{ing}	CR_{inh}	CR_{dermal}	CR
As	Adults	4.51×10^{-2}	3.36×10^{-4}	4.35×10^{-4}	4.59×10^{-2}	2.03×10^{-5}	2.18×10^{-8}	8.09×10^{-8}	2.04×10^{-5}
	Children	3.02×10^{-1}	5.83×10^{-4}	2.05×10^{-3}	3.05×10^{-1}	1.36×10^{-4}	3.78×10^{-8}	3.81×10^{-7}	1.36×10^{-4}
Cd	Adults	5.32×10^{-4}	1.98×10^{-5}	8.49×10^{-5}	6.42×10^{-4}	3.27×10^{-6}	3.57×10^{-10}		3.27×10^{-6}
	Children	3.57×10^{-3}	3.44×10^{-5}	3.99×10^{-4}	4.03×10^{-3}	2.19×10^{-5}	6.19×10^{-10}		2.19×10^{-5}
Cr	Adults	3.63×10^{-2}	4.06×10^{-4}	1.45×10^{-2}	5.12×10^{-2}	5.45×10^{-5}	4.87×10^{-9}	8.68×10^{-6}	6.32×10^{-5}
	Children	2.43×10^{-1}	7.04×10^{-4}	6.81×10^{-2}	3.12×10^{-1}	3.66×10^{-4}	8.45×10^{-9}	4.09×10^{-5}	4.06×10^{-4}
Hg	Adults	6.36×10^{-4}		3.61×10^{-5}	6.82×10^{-4}				
	Children	4.26×10^{-3}		1.70×10^{-4}	4.50×10^{-3}				
Pb	Adults	2.99×10^{-2}		3.19×10^{-4}	3.03×10^{-2}	3.56×10^{-7}	1.88×10^{-10}		3.56×10^{-7}
	Children	2.01×10^{-1}		1.50×10^{-3}	2.02×10^{-1}	2.39×10^{-6}	3.26×10^{-10}		2.39×10^{-6}
THI/TCR	Adults	1.12×10^{-1}	7.62×10^{-4}	1.54×10^{-2}	1.29×10^{-1}	7.52×10^{-5}	2.72×10^{-8}	8.77×10^{-6}	8.40×10^{-5}
	Children	7.54×10^{-1}	1.32×10^{-3}	7.22×10^{-2}	8.28×10^{-1}	5.04×10^{-4}	4.72×10^{-8}	4.12×10^{-5}	5.45×10^{-4}

The kriging technique was further employed to map out the spatial patterns of THI and TCR among both adults and children (Figure 5). The high TCR values of adults and children were primarily located in the southwest and central parts areas (Figure 5a,c), and the high THI values were mainly distributed in the southwest, central and northeast areas (Figure 5b,d).

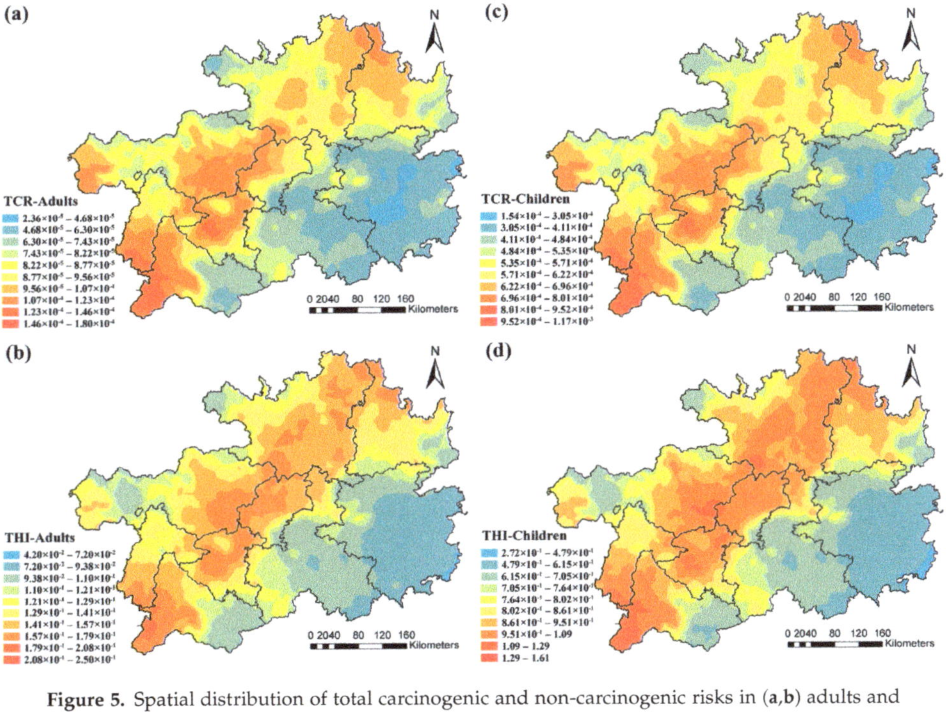

Figure 5. Spatial distribution of total carcinogenic and non-carcinogenic risks in (**a**,**b**) adults and (**c**,**d**) children.

3.4. Probabilistic Risk Assessment

The probabilistic assessment by Monte Carlo simulation was shown in Figures 6 and 7. The probabilistic risks ranked as follows: for mean HI values, Cr(5.18×10^{-2}) > As(4.64×10^{-2}) > Pb(3.06×10^{-2}) > Hg(6.79×10^{-4}) > Cd(6.45×10^{-4}) in adults (Figure 6b–f) and Cr(3.15×10^{-1}) > As(3.08×10^{-1}) > Pb(2.04×10^{-1}) > Hg(4.47×10^{-3}) > Cd(4.04×10^{-3}) in children (Figure 6h–l); similarly, for mean CR values, the order was Cr(6.40×10^{-5}) > As(2.07×10^{-5}) > Cd(3.29×10^{-6}) > Pb(3.61×10^{-7}) in adults

(Figure 7b–e) and the order was Cr(4.11×10^{-4}) > As(1.38×10^{-4}) > Cd(2.20×10^{-5}) > Pb(2.41×10^{-6}) in children (Figure 7g–j). Obviously, the Cr values both in CR and HI were the highest among all of the investigated HMs. In adults, HI(Cr) contributed 39.85% to THI, followed by HI(As) with 35.69%. In children, HI(Cr) contributed 37.68% to THI, followed by HI(As) with 36.84%. In children, there was an 11.67% probability that the THI value exceeded the exposure risk value. The acceptable threshold for CR(Cr) was likely to be exceeded in adults, but it was guaranteed to be exceeded in children, with a 100% probability. Additionally, there was a 94.62% probability that the acceptable threshold for CR(As) would be exceeded in children. The total probabilistic carcinogenic risk was 20.69% for adults and 100% for children, respectively. The Student's t-test showed that there were significant differences in all health risk indicators between children and adults (Figures S3 and S4). For THI, TCR, As, Cd, Cr, Hg and Pb, the probabilistic non-carcinogenic and carcinogenic risks were notably greater in children compared to adults ($p < 0.0001$).

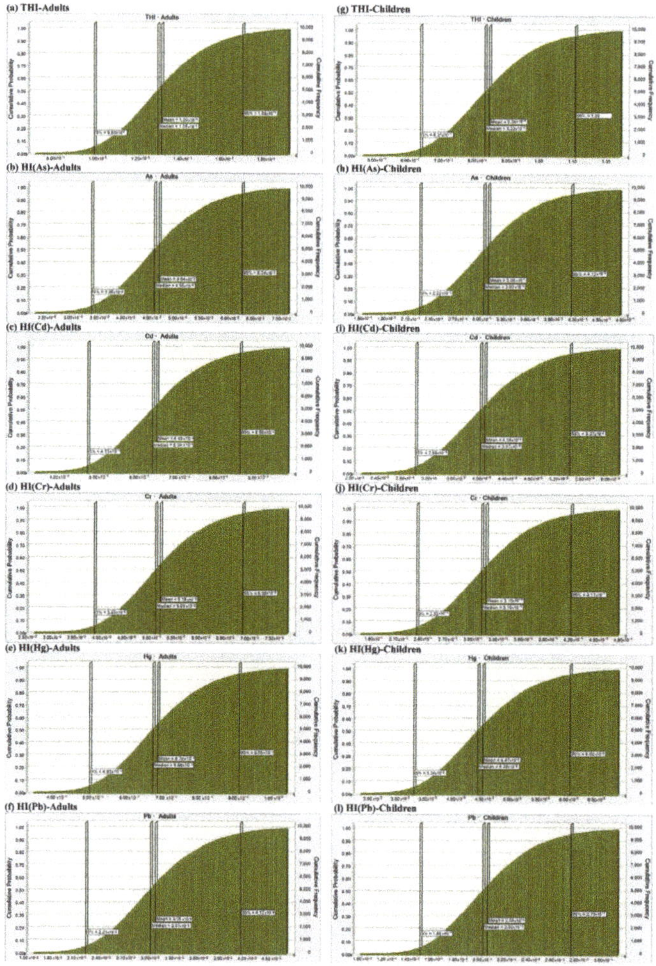

Figure 6. Probabilistic non-carcinogenic risk assessment of heavy metals in (**a**–**f**) adults and (**g**–**l**) children.

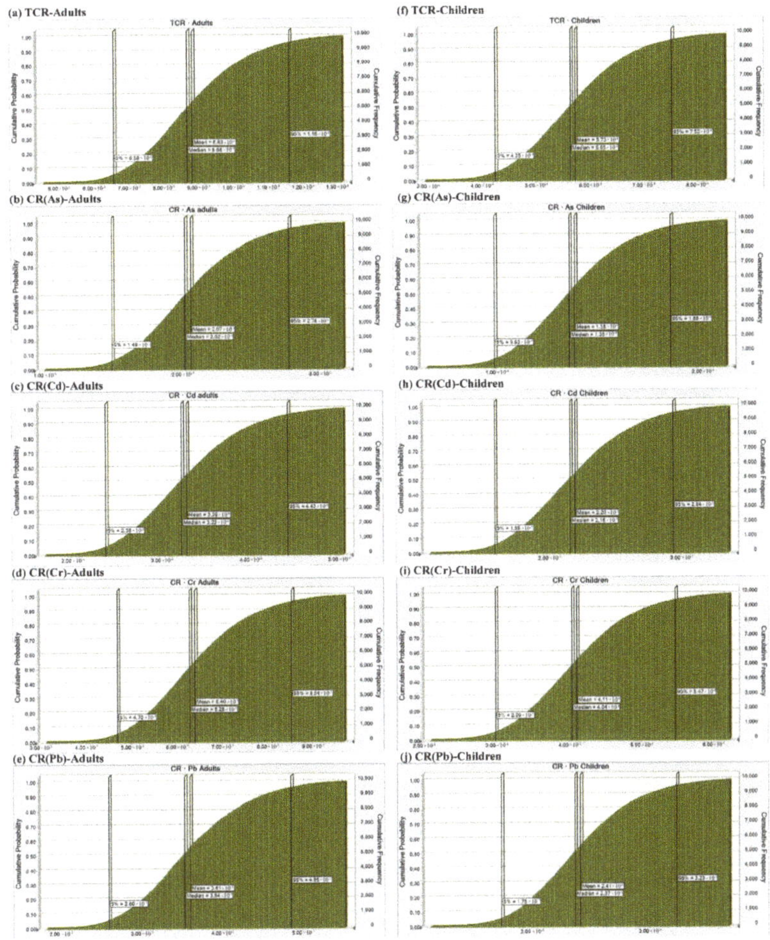

Figure 7. Probabilistic carcinogenic risk assessment of heavy metals in (**a–e**) adults and (**f–j**) children.

3.5. Health Risk Assessment Based on the TTD Method

Oral ingestion, as the primary exposure way of health risk, was selected as a non-carcinogenic risk assessment modified by the TTD. From the target organs, the cumulative risks in our study were 0.09, 0.08, 0.07, 0.06 and 0.2 in adults (Figure 8f) and 0.61, 0.55, 0.52, 0.46 and 0.17 in children (Figure 8g), respectively. It showed that HI_{TTD}(Hematological) had the highest contribution rate to THI_{TTD}, which was 27.27% and 26.41% in adults and children, respectively. From the perspective, the Pb, As, Cr and Cd values for cumulative risks were 0.14, 0.10, 0.08 and 0.005 in adults and 1, 0.75, 0.53, and 0.03 in children, respectively. The data showed that the contributions of HI_{TTD}(Pb) were 42.42% in adults and 43.29% in children. In total, The THI_{TTD} was 2.56 times higher than the definitive risk assessment of THI_{Adults} and 2.79 times that of $THI_{Children}$. The health risks of HMs in different target organs were illustrated in Figure 8. The neurological and cardiovascular systems were most sensitive to As (Figure 8a,c), the renal system to Pb (Figure 9b), and the hematological and testicular systems to Cr (Figure 8d,e). In addition, children faced greater health risks compared to adults.

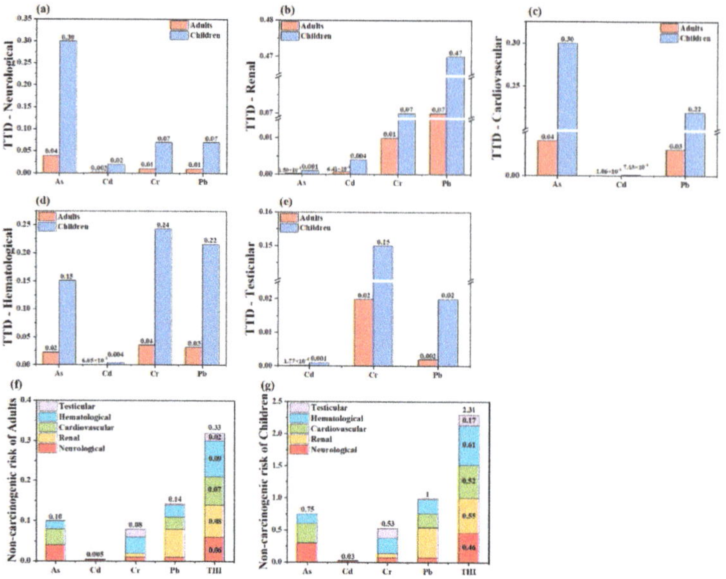

Figure 8. Non-carcinogenic risk of (**a–e**) different target organs and (**f**,**g**) different populations based on TTD.

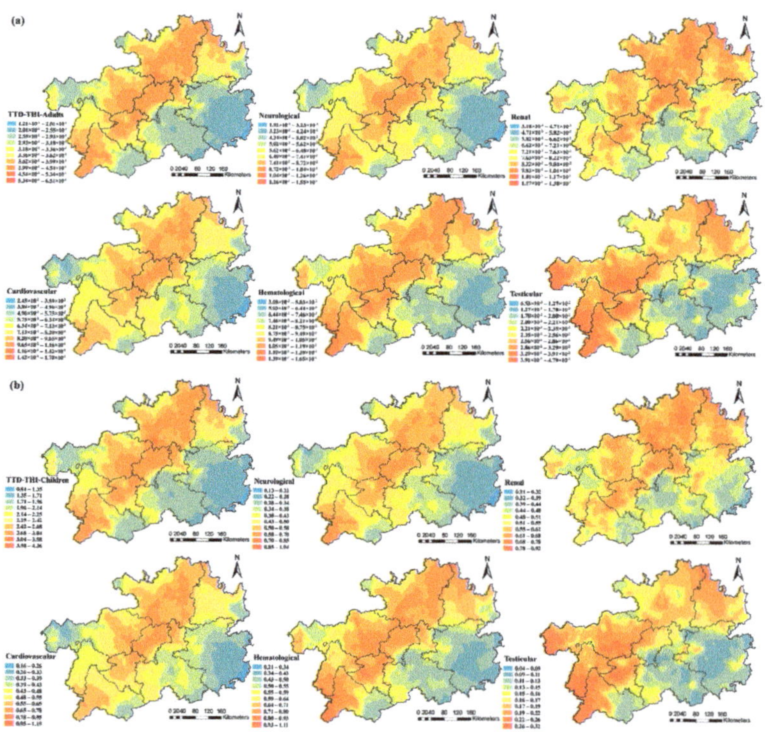

Figure 9. Spatial distribution of total non-carcinogenic risks of different target organs in adults (**a**) and children (**b**).

The spatial pattern of HI_{TTD} was uniformly comparable across adults (Figure 9a) and children (Figure 9b) for various target organs. The high-risk areas for the neurological and cardiovascular systems were primarily in the central region, those for the renal system in the northern, and those for the testicular system in the west. The high values of THI_{TTD} were primarily located in the southwest, central and northeast regions.

4. Discussion

4.1. Heavy Metals Pollution Analysis

In most cases, the adsorption of metal elements onto soil particle surfaces intensifies as soil pH levels increase [60]. We found that the higher Cd concentrations of samples are higher at the low pH. To some extent, it might be contributed to the low solubility of soil Cd at a high pH accounting with the properties of calcareous soil [61]. In contrast, the present study indicated that As exceeded the limit at high pH levels. As previously study described, the solubility of As in soil is possibly rising with the soil pH increasing [62]. Furthermore, the CV reflected the variability and dispersion of soil elements. Elements with high CV may be affected by human activities [63]. As, Cd, Cr and Hg showed high variability (CV $\geq$ 36%), indicating that they had high spatial heterogeneity [64], which may be due to the impacts of parent rocks and the process of soil formation [16].

The PI of Cd was the highest, primarily due to its relatively high toxicological response factor [65]. Investigation of the agricultural soil near the mining area in central Guizhou showed that the P_N was 2.5 [14], which was higher than the result of this study, as it was near the mining area possibly. Although the PI for other HMs were within the standard limits, 24.46% of the sites showed Cd contamination. Therefore, the P_N was 1.77, indicating that Guizhou province was under slight pollution. The P_N describes the possibility of pollution, the risk amount of the indicated pollution and it is also able to measure the reach of HMs pollution to the topsoil level, taking into account the risks of all referenced HMs [66]. The Igeo index considers the effects of natural diagenesis and anthropogenic activities [67], and the Igeo of 31.62% of Hg was slightly polluted, indicating that Hg was affected by anthropogenic activities. Guizhou is one of the major Hg-producing regions in China, consistently ranking first in terms of Hg ore reserves and production. The related mining and smelting activities will produce a large amount of Hg-containing waste gas, waste water and waste residue, which is very likely to lead to HMs contamination of its neighboring soils [68]. Therefore, the contamination of Hg in farmland soil should be noted within the Guizhou Province. Moreover, the Er of 10.95% Cd indicated a moderate pollution level. Although the results of RI indicated that the study area was at a low risk level, the impact of Cd should still be taken seriously. The Igeo focuses on quantifying the effects of anthropogenic pollution, while other indices (such as Er and RI) pay more attention to toxicity responses or comprehensive risk assessment. Therefore, PI and Er described the pollution risk level of Cd, while Igeo indicated the impact of human activities on Hg. The pollution of Cd and Hg in the farmland soil of Guizhou Province deserves attention, suggesting that priority should be given to control and reduce their risk to the environment. The deficiency of this study is that only five HMs were investigated. However, the influence of other HMs (such as Cu, Co, Zn, etc.) is also very important. Therefore, in the subsequent research work, we consider including these several HMs in the study to assess the soil environmental quality more comprehensively.

4.2. Deterministic and Probabilistic Risk Assessment

Deterministic health risk assessment was used to estimate carcinogenic and non-carcinogenic risks for adults and children via ingestion, inhalation, and skin contact. Kyere et al. [69] and Demirtepe et al. [70] showed that higher HI levels were typically observed in the ingestion routes compared to inhalation and skin contact routes. This study also showed that the ingestion route contributed the most to HI. Therefore, this pathway should be valued. Similar to previous studies [71,72], our study indicated that children exhibited higher non-carcinogenic and carcinogenic risks than adults ($p < 0.0001$), primarily owing to their conduct, biological traits, and exposure duration [73]. In addition, Lu et al. [74] revealed that As and Cr contributed high CR was noteworthy, particularly in the southwestern region, which aligned with the findings of this study. The CR of Cr was higher, potentially due to its lower slope factor, posing a greater carcinogenic risk than other HMs. Meantime, there are many mineral resources in Guizhou Province with high concentrations of Cr [75]. Therefore, the health risks posed by As and Cr to children deserve attention in the southwestern region.

Monte Carlo simulation, a widely used probabilistic risk assessment technique, minimizes uncertainty and offers more comprehensive results during risk evaluation [22]. Through the evaluation of two methods, Cr had the highest risk in HI and CR, followed by As; and CR values for Cr and As exceeded the acceptable range in children. However, the mean HI and CR values of As, Cd, Cr, and Pb, for the probabilistic approach were slightly higher than those for the deterministic approach. Previous research used deterministic values to assess health risks, which eventually may underestimate the risk outcomes [76,77]. A significant reduction has been observed in exceedance probabilities for THI for children and adult TCR and CR(As) for adults relative to their safety thresholds. These phenomena suggested that the above indicators may be overestimated in deterministic assessments. Additionally, the CR of As for children exceeded the probability risk increased, among from 57.30% in deterministic assessment to 94.62%, which indicated that the risk may be underestimated. To our knowledge, deterministic and probabilistic techniques are widely used to estimate human health risks posed by various pollutants [78]. Combining the two methods to explore the health risks of HMs to people can provide a scientific foundation for policymakers to achieve risk management. A key limitation is the absence of formal sensitivity analysis. Although the main purpose of this study was to assess the overall risk probability range of soil heavy metals to the population, and Monte Carlo simulation effectively quantified the variability of the results under the uncertainty of the given parameters, this limited the identification of key risk drivers. Future studies should incorporate sensitivity and uncertainty analysis to further explain the research results in depth. In addition, Jin et al. [79] estimated the health risks of HMs in food, which gave us great enlightenment. Subsequently, we consider conducting an assessment of HMs pollution and health risks related to soil, crops and humans.

4.3. Health Risk Assessment Modified by the TTD Method

The health risks posed by HMs in soil to local populace were objectively evaluated by both the deterministic and probabilistic risks. However, traditional health risk assessment models only consider the most sensitive effect target organ of HMs, while actual risks arise from damage to multiple target organs simultaneously. This may result in an underestimation of the non-carcinogenic health risks posed by soil HMs contamination to humans [57,80]. Therefore, TTD was further used for risk assessment of specific target organs.

On account of Pb being more sensitive to the toxic effects of target organs, its cumulative risk value was higher than other HMs. More importantly, the HI_{TTD} value of Pb in children was 1. Therefore, the health risks caused by Pb to children should be paid attention to. In terms of a single target organ, the accumulation risk in the hematology was the highest, indicating that HMs in agricultural soil might cause damage to the hematological system of the population, although it is still within the safe threshold of soil risk.

Compared with the traditional deterministic assessment results, the HI_{TTD} evaluated by the TTD model was 1.56~7.79 times than that of HI, which was mainly due to the joint action of two or more HMs in these target organs [81]. Thus, utilizing the TTD model allows for more precise measurement and comparison of cumulative risks associated with various HMs against risk values obtained through traditional methods, providing a deeper understanding of health risks to target organs [81]. To this end, the TTD addresses the limitation of the traditional method in comprehensively assessing risks across multiple target organs.

In addition, consistent with the results of traditional methods, the health risks in the southwest region are noteworthy. Due to the lack of supporting data and uneven medical resource allocation, it cannot be determined whether the incidence rates of relevant diseases in various regions of Guizhou align with research findings. Therefore, residents in high-risk areas should be mindful of undergoing regular health checks for related target organ diseases. However, none of these three health risk assessment techniques take into account the bioavailability of HMs in the soil, which may lead to an overestimation of health risks. This is the deficiency of this study. Future research will focus on the bioavailability of soil HMs and animal experiments to fully explore and verify the carcinogenic and non-carcinogenic health risks brought by related metal pollutants to the human body, to evaluate the health risks of soil HMs more comprehensively. It will facilitate decision-making regarding health risks and the formulation of appropriate public health measures.

5. Conclusions

Cd and Hg were the main pollutants in agricultural soil of Guizhou Province, and their distribution was affected by industrial activities and soil pH. The P_N indicated slight pollution in the farmland soil within the province. Children were at higher risk of non-carcinogenic and carcinogenic than adults, and the CR of Cr (3.12×10^{-1}) and As (3.05×10^{-1}) was particularly prominent, which exceeded the acceptable range. However, the carcinogenic risk of adults was less than 1×10^{-4} and did not exceed the standard. Convergent findings from deterministic modeling and Monte Carlo simulations revealed CR for As and Cr exceeding safety thresholds in children. Further, the TTD was used to assess multi-organ risk, revealing a higher risk of soil HMs for hematological health, with notable health risks posed by Pb in children. This approach addresses the limitation of the traditional method in comprehensively assessing risks across multiple target organs. It is noted that spatial distribution analysis suggested that the southwestern region of Guizhou Province should be prioritized for health risk management and control. By integrating the uniqueness of geological environments, multi-dimensional health risk assessments, and spatial distributions, this study provides a scientific basis for assessing HMs pollution risks and soil health risks in karst regions.

Supplementary Materials: The following supporting information can be downloaded at: https://www.mdpi.com/article/10.3390/toxics13060515/s1, Figure S1. Non-carcinogenic risk significance test analysis for adults and children Note: **** $p < 0.0001$; Figure S2. Carcinogenic risk significance test analysis for adults and children. Note: **** $p < 0.0001$; Figure S3. Probabilistic non-carcinogenic risk significance test analysis for adults and children. Note: **** $p < 0.0001$; Figure S4. Probabilistic carcinogenic risk significance test analysis for adults and children. Note: **** $p < 0.0001$; Table S1: Classification of the PI and P_N; Table S2: Classification of the Igeo; Table S3: Classification of the Er and RI; Table S4: Description of parameters in health risk assessment; Table S5: Reference dose (RfD) and cancer slope factor (SF) via oral (Ingestion), inhalation and dermal contact; Table S6: Distribution settings for each parameter in the Monte Carlo simulation; Table S7: The corresponding target organ toxicity doses for each heavy metal.

Author Contributions: Q.L.: Conceptualization, methodology, validation and writing—original draft preparation; D.L.: Conceptualization, validation and writing—review and editing; Z.W.: Validation and formal analysis; D.S.: Data curation and funding acquisition; T.Z.: Writing—review and editing, supervision and project administration; Q.Z.: Resources, project administration and funding acquisition. All authors have read and agreed to the published version of the manuscript.

Funding: This research was funded by the National Natural Science Foundation of China [42467054; 42167054]; Guizhou Provincial Foundation for Excellent Scholars Program (No. GCC[2023]076); Guizhou Provincial Basic Research Program (Natural Science) ([2023]317); Guizhou Provincial Key Technology R&D Program (No. 2024078); the Open Foundation for Key Laboratory of Environmental Pollution Monitoring and Disease Control, Ministry of Education ([2022]441); (GMU-2023-HJZ-03); The Central Government Supports the Reform and Development of Local Colleges and Universities ([2023]067).

Institutional Review Board Statement: Not applicable.

Informed Consent Statement: Not applicable.

Data Availability Statement: The data presented in this study are available on request from the corresponding author.

Conflicts of Interest: The authors declare no conflicts of interest.

References

1. Su, C.; Wang, J.; Chen, Z.; Meng, J.; Yin, G.; Zhou, Y.; Wang, T. Sources and health risks of heavy metals in soils and vegetables from intensive human intervention areas in South China. *Sci. Total Environ.* **2023**, *857*, 159389. [CrossRef] [PubMed]
2. Vareda, J.P.; Valente, A.J.M.; Durães, L. Assessment of heavy metal pollution from anthropogenic activities and remediation strategies: A review. *J. Environ. Manag.* **2019**, *246*, 101–118. [CrossRef]
3. Wei, L.; Wang, K.; Noguera, D.R.; Jiang, J.; Oyserman, B.; Zhao, N.; Zhao, Q.; Cui, F. Transformation and speciation of typical heavy metals in soil aquifer treatment system during long time recharging with secondary effluent: Depth distribution and combination. *Chemosphere* **2016**, *165*, 100–109. [CrossRef] [PubMed]
4. Konyshev, A.A.; Sidkina, E.S.; Cherkasova, E.V.; Mironenko, M.V.; Gridasov, A.G.; Zhilkina, A.V.; Bugaev, I.A. Migration Forms of Heavy Metals and Chemical Composition of Surface Waters in the "Arsenic" Shaft Area (Pitkäranta Ore District, South Karelia). *Geochem. Int.* **2020**, *58*, 1068–1074. [CrossRef]
5. Shi, H.; Li, J. Research on Heavy Metal Pollution and Comprehensive Treatment of Farmland Soil. *Front. Chem. Sci. Eng.* **2022**, *2*, 8–12. [CrossRef]
6. Xu, D.; Fu, R.; Liu, H.; Guo, X. Current knowledge from heavy metal pollution in Chinese smelter contaminated soils, health risk implications and associated remediation progress in recent decades: A critical review. *J. Clean. Prod.* **2021**, *286*, 124989. [CrossRef]
7. Chen, H.; Teng, Y.; Lu, S.; Wang, Y.; Wang, J. Contamination features and health risk of soil heavy metals in China. *Sci. Total Environ.* **2015**, *512–513*, 143–153. [CrossRef]
8. Yang, Q.; Yang, Z.; Zhang, Q.; Liu, X.; Zhuo, X.; Wu, T.; Wang, L.; Wei, X.; Ji, J. Ecological risk assessment of Cd and other heavy metals in soil-rice system in the karst areas with high geochemical background of Guangxi, China. *Sci. China Earth Sci.* **2021**, *64*, 1126–1139. [CrossRef]
9. Wen, Y.; Li, W.; Yang, Z.; Zhang, Q.; Ji, J. Enrichment and source identification of Cd and other heavy metals in soils with high geochemical background in the karst region, Southwestern China. *Chemosphere* **2020**, *245*, 125620. [CrossRef]
10. Xia, X.; Ji, J.; Yang, Z.; Han, H.; Huang, C.; Li, Y.; Zhang, W. Cadmium risk in the soil-plant system caused by weathering of carbonate bedrock. *Chemosphere* **2020**, *254*, 126799. [CrossRef]
11. Luo, K.; Liu, H.; Liu, Q.; Tu, Y.; Yu, E.; Xing, D. Cadmium accumulation and migration of 3 peppers varieties in yellow and limestone soils under geochemical anomaly. *Environ. Technol.* **2022**, *43*, 10–20. [CrossRef] [PubMed]
12. Pu, W.; Sun, J.; Zhang, F.; Wen, X.; Liu, W.; Huang, C. Effects of copper mining on heavy metal contamination in a rice agrosystem in the Xiaojiang River Basin, southwest China. *Acta Geochim.* **2019**, *38*, 753–773. [CrossRef]
13. Guo, Y.; Wu, R.; Guo, C.; Lv, J.; Wu, L.; Xu, J. Occurrence, sources and risk of heavy metals in soil from a typical antimony mining area in Guizhou Province, China. *Environ. Geochem. Health* **2022**, *45*, 3637–3651. [CrossRef]
14. Cui, W.; Mei, Y.; Liu, S.; Zhang, X. Health risk assessment of heavy metal pollution and its sources in agricultural soils near Hongfeng Lake in the mining area of Guizhou Province, China. *Front. Public Health* **2023**, *11*, 1276925. [CrossRef] [PubMed]
15. Tang, M.; Lu, G.; Fan, B.; Xiang, W.; Bao, Z. Bioaccumulation and risk assessment of heavy metals in soil-crop systems in Liujiang karst area, Southwestern China. *Environ. Sci. Pollut. Res.* **2020**, *28*, 9657–9669. [CrossRef]

16. Qin, Y.; Zhang, F.; Xue, S.; Ma, T.; Yu, L. Heavy Metal Pollution and Source Contributions in Agricultural Soils Developed from Karst Landform in the Southwestern Region of China. *Toxics* **2022**, *10*, 568. [CrossRef]
17. Adimalla, N.; Chen, J.; Qian, H. Spatial characteristics of heavy metal contamination and potential human health risk assessment of urban soils: A case study from an urban region of South India. *Ecotoxicol. Environ. Saf.* **2020**, *194*, 110406. [CrossRef]
18. Kampa, M.; Castanas, E. Human health effects of air pollution. *Environ. Pollut.* **2008**, *151*, 362–367. [CrossRef]
19. Wang, Y.; Duan, X.; Wang, L. Spatial distribution and source analysis of heavy metals in soils influenced by industrial enterprise distribution: Case study in Jiangsu Province. *Sci. Total Environ.* **2020**, *710*, 134953. [CrossRef]
20. Chen, R.; Zhang, Q.; Chen, H.; Yue, W.; Teng, Y. Source apportionment of heavy metals in sediments and soils in an interconnected river-soil system based on a composite fingerprint screening approach. *J. Hazard. Mater.* **2021**, *411*, 125125. [CrossRef]
21. Huang, J.; Wu, Y.; Sun, J.; Li, X.; Geng, X.; Zhao, M.; Sun, T.; Fan, Z. Health risk assessment of heavy metal(loid)s in park soils of the largest megacity in China by using Monte Carlo simulation coupled with Positive matrix factorization model. *J. Hazard. Mater.* **2021**, *415*, 125629. [CrossRef] [PubMed]
22. Chen, H.; Wang, L.; Hu, B.; Xu, J.; Liu, X. Potential driving forces and probabilistic health risks of heavy metal accumulation in the soils from an e-waste area, southeast China. *Chemosphere* **2022**, *289*, 133182. [CrossRef] [PubMed]
23. Yang, S.; Liu, X.; Xu, J. New Perspectives about Health Risk Assessment of Soil Heavy Metal Pollution-Origin and Prospects of Probabilistic Risk Analysis. *Acta Pedol. Sin.* **2022**, *59*, 28–37. [CrossRef]
24. Yang, B.; Li, W.; Xiong, J.; Yang, J.; Huang, R.; Xie, P. Health Risk Assessment of Heavy Metals in Soil of Lalu Wetland Based on Monte Carlo Simulation and ACPS-MLR. *Water* **2023**, *15*, 4223. [CrossRef]
25. Tudi, M.; Li, H.; Li, H.; Wang, L.; Lyu, J.; Yang, L.; Tong, S.; Yu, Q.J.; Ruan, H.D.; Atabila, A.; et al. Exposure Routes and Health Risks Associated with Pesticide Application. *Toxics* **2022**, *10*, 335. [CrossRef]
26. Ding, D.; Kong, L.; Jiang, D.; Wei, J.; Cao, S.; Li, X.; Zheng, L.; Deng, S. Source apportionment and health risk assessment of chemicals of concern in soil, water and sediment at a large strontium slag pile area. *J. Environ. Manag.* **2022**, *304*, 114228. [CrossRef]
27. Eslami, Z.; Mahdavi, V.; Tajdar-oranj, B. Probabilistic health risk assessment based on Monte Carlo simulation for pesticide residues in date fruits of Iran. *Environ. Sci. Pollut. Res.* **2021**, *28*, 42037–42050. [CrossRef]
28. Peng, C.; Cai, Y.; Wang, T.; Xiao, R.; Chen, W. Regional probabilistic risk assessment of heavy metals in different environmental media and land uses: An urbanization-affected drinking water supply area. *Sci. Rep.* **2016**, *6*, 37084. [CrossRef]
29. Wilbur, S.B.; Hansen, H.; Pohl, H.; Colman, J.; McClure, P. Using the ATSDR Guidance Manual for the Assessment of Joint Toxic Action of Chemical Mixtures. *Environ. Toxicol. Pharmacol.* **2004**, *18*, 223–230. [CrossRef]
30. Kong, X.; Liu, T.; Yu, Z.; Chen, Z.; Lei, D.; Wang, Z.; Zhang, H.; Li, Q.; Zhang, S. Heavy Metal Bioaccumulation in Rice from a High Geological Background Area in Guizhou Province, China. *Int. J. Environ. Res. Public Health* **2018**, *15*, 2281. [CrossRef]
31. Ministry of Land and Resources of the People's Republic of China. *Specification of Land Quality Geochemical Assessment*; Ministry of Land and Resources of the People's Republic of China: Beijing, China, 2016.
32. Ministry of Environmental Protection. *The Technical Specification for Soil Environmental Monitoring*; Ministry of Environmental Protection: Beijing, China, 2004.
33. Ministry of Environmental Protection. *Soil and Sediment—Digestion of Total Metal Elements—Microwave Assisted Acid Digestion Method*; Ministry of Environmental Protection: Beijing, China, 2017.
34. He, S.; Niu, Y.; Xing, L.; Liang, Z.; Song, X.; Ding, M.; Huang, W. Research progress of the detection and analysis methods of heavy metals in plants. *Front. Plant Sci.* **2024**, *15*, 1310328. [CrossRef] [PubMed]
35. General Administration of Quality Supervision, Inspection and Quarantine. *Soil Quality—Determination of Total Mercury, Total Arsenic, Total Lead—Atomic Fluorescence Part I: Determination of Total Mercury in Soil*; General Administration of Quality Supervision, Inspection and Quarantine: Beijing, China, 2008.
36. General Administration of Quality Supervision, Inspection and Quarantine. *Soil Quality—Determination of Total Mercury, Total Arsenic, Total Lead—Atomic Fluorescence Part II: Determination of Total Arsenic in Soil*; General Administration of Quality Supervision, Inspection and Quarantine: Beijing, China, 2008.
37. Wang, T.; Yang, Y.; Ya, Y.; Mo, L.; Fan, Y.; Liao, J.; Huang, D.; Tan, H. Determination of Arsenic and Mercury in Soil by Microwave Digestion and Hidride GenerationAtomic Fluorescence Spectrometry. *ACS Agric. Sci. Technol.* **2013**, *14*, 651–653.
38. Hakanson, L. An ecological risk index for aquatic pollution control. A sedimentological approach. *Water Res.* **1980**, *14*, 975–1001. [CrossRef]
39. Nemerow, N.L. *Stream, Lake, Estuary, and Ocean Pollution*; Van Nostrand Reinhold: New York, NY, USA, 1985.
40. Ministry of Environmental Protection. *Soil Environmental Quality Risk Control Standard for Soil Contamination of Agricultural Land*; Ministry of Environmental Protection: Beijing, China, 2018.
41. Förstner, U.; Ahlf, W.; Calmano, W.; Kersten, M. Sediment Criteria Development. In *Sediments and Environmental Geochemistry*; Springer: Berlin/Heidelberg, Germany, 1990; pp. 311–338. [CrossRef]
42. Müller, G. Index of geoaccumulation in sediments of the Rhine River. *GeoJournal* **1969**, *2*, 108–118.

43. CNEMC. *The Backgrounds of Soil Environment of Guizhou, China*; China National Environmental Monitoring Center: Beijing, China, 1990.
44. Loska, K.; Wiechula, D.; Korus, I. Metal contamination of farming soils affected by industry. *Environ. Int.* **2004**, *30*, 159–165. [CrossRef]
45. USEPA; Exposure Analysis and Risk Characterization Group; Moya, J. *Exposure Factors Handbook*; USEPA National Center for Environmental Assessment: Washington, DC, USA, 1997.
46. USEPA. *Supplemental Guidance for Developing Soil Screening Levels for Superfund Sites*; Office of Emergency and Remedial Response: Washington, DC, USA, 2002.
47. Ministry of Environmental Protection. *Chinese Population Exposure Parameter Manual*; Ministry of Environmental Protection: Beijing, China, 2013.
48. Liu, H.; Wang, H.; Zhang, Y.; Yuan, J.; Peng, Y.; Li, X.; Shi, Y.; He, K.; Zhang, Q. Risk assessment, spatial distribution, and source apportionment of heavy metals in Chinese surface soils from a typically tobacco cultivated area. *Environ. Sci. Pollut. Res. Int.* **2018**, *25*, 16852–16863. [CrossRef]
49. MohseniBandpi, A.; Eslami, A.; Ghaderpoori, M.; Shahsavani, A.; Jeihooni, A.K.; Ghaderpoury, A.; Alinejad, A. Health risk assessment of heavy metals on PM2.5 in Tehran air, Iran. *Data Brief* **2018**, *17*, 347–355. [CrossRef]
50. USEPA. *Risk Assessment Guidance for Superfund (RAGS)*; U.S. Environment Protection Agency: Washington, DC, USA, 2009.
51. Sarwar, T.; Shahid, M.; Natasha, N.; Khalid, S.; Haidar Shah, A.; Ahmad, N.; Naeem, M.A.; Khan, Z.U.H.; Murtaza, B.; Bakhat, H. Quantification and risk assessment of heavy metal build-up in soil–plant system after irrigation with untreated city wastewater in Vehari, Pakistan. *Environ. Geochem. Health* **2020**, *42*, 4281–4297. [CrossRef]
52. Yang, X.; Cheng, B.; Gao, Y.; Zhang, H.; Liu, L. Heavy metal contamination assessment and probabilistic health risks in soil and maize near coal mines. *Front. Public Health* **2022**, *10*, 1004579. [CrossRef]
53. Liu, Z.; Du, Q.; Guan, Q.; Luo, H.; Shan, Y.; Shao, W. A Monte Carlo simulation-based health risk assessment of heavy metals in soils of an oasis agricultural region in northwest China. *Sci. Total Environ.* **2023**, *857*, 159543. [CrossRef]
54. Duan, X.; Zhao, X.; Wang, B.; Chen, Y.; Cao, S. *Highlights of the Chinese Exposure Factors Handbook (Adults)*; Academic Press: Cambridge, MA, USA, 2015.
55. Chen, R.; Chen, H.; Song, L.; Yao, Z.; Meng, F.; Teng, Y. Characterization and source apportionment of heavy metals in the sediments of Lake Tai (China) and its surrounding soils. *Sci. Total Environ.* **2019**, *694*, 133819. [CrossRef] [PubMed]
56. USEPA. *Exposure Factors Handbook: 2011 Edition*; National Center for Environmental Assessment Office of Research and Development: Washington, DC, USA, 2011.
57. Liu, L.; Han, J.; Qian, Y.; Zhang, Z.; Guo, S. Assessment of heavy metal non-carcinogenic health risk in solidified fly ash using TTD and WOE methods. *Environ. Chem.* **2019**, *38*, 1014–1020. (In Chinese)
58. ATSDR. *Guidance Manual for the Assessment of Joint Toxic Action of Chemical Mixtures*; ATSDR: Atlanta, GA, USA, 2001.
59. ATSDR. *Interaction Profile for Arsenic, Cadmium, Chromium and Lead [Online]*; ATSDR: Atlanta, GA, USA, 2004.
60. Vega, F.A.; Covelo, E.F.; Andrade, M.L. Competitive sorption and desorption of heavy metals in mine soils: Influence of mine soil characteristics. *J. Colloid. Interface Sci.* **2006**, *298*, 582–592. [CrossRef]
61. Zhang, S.; Song, J.; Cheng, Y.; McBride, M.B. Derivation of regional risk screening values and intervention values for cadmium-contaminated agricultural land in the Guizhou Plateau. *Land Degrad. Dev.* **2018**, *29*, 2366–2377. [CrossRef]
62. Wang, X.; Jia, Y.; Jiang, R. Effects of As pollutant in soil on crop growth and safety of agricultural products under Cd stress. *Ecol. Environ. Sci.* **2009**, *18*, 2132–2136. (In Chinese)
63. Yang, S.; Qu, Y.; Ma, J.; Liu, L.; Wu, H.; Liu, Q.; Gong, Y.; Chen, Y.; Wu, Y. Comparison of the concentrations, sources, and distributions of heavy metal(loid)s in agricultural soils of two provinces in the Yangtze River Delta, China. *Environ. Pollut.* **2020**, *264*, 114688. [CrossRef]
64. Zhu, Y.; An, Y.; Li, X.; Cheng, L.; Lv, S. Geochemical characteristics and health risks of heavy metals in agricultural soils and crops from a coal mining area in Anhui province, China. *Environ. Res.* **2024**, *241*, 117670. [CrossRef]
65. Obiri-Nyarko, F.; Duah, A.A.; Karikari, A.Y.; Agyekum, W.A.; Manu, E.; Tagoe, R. Assessment of heavy metal contamination in soils at the Kpone landfill site, Ghana: Implication for ecological and health risk assessment. *Chemosphere* **2021**, *282*, 131007. [CrossRef]
66. Yari, A.A.; Varvani, J.; Zare, R. Assessment and zoning of environmental hazard of heavy metals using the Nemerow integrated pollution index in the vineyards of Malayer city. *Acta Geophys.* **2020**, *69*, 149–159. [CrossRef]
67. Zhao, L.; Xu, Y.; Hou, H.; Shangguan, Y.; Li, F. Source identification and health risk assessment of metals in urban soils around the Tanggu chemical industrial district, Tianjin, China. *Sci. Total Environ.* **2014**, *468–469*, 654–662. [CrossRef]
68. Ma, L.; Zhou, L.; Song, B.; Wang, F.; Zhang, Y.; Wu, Y. Mercury Pollution in Dryland Soil and Evaluation of Maize Safety Production in Guizhou Province. *Environ. Sci.* **2023**, *44*, 2868–2878. (In Chinese)
69. Kyere, V.N.; Greve, K.; Atiemo, S.M.; Amoako, D.; Aboh, I.J.K.; Cheabu, B.S. Contamination and Health Risk Assessment of Exposure to Heavy Metals in Soils from Informal E-Waste Recycling Site in Ghana. *Emerg. Sci. J.* **2018**, *2*, 428. [CrossRef]

70. Demirtepe, H. Soil Contamination by Metals/Metalloids around an Industrial Region and Associated Human Health Risk Assessment. *J. Adv. Res. Nat. Appl. Sci.* **2024**, *10*, 91–105. [CrossRef]
71. Ahmad, W.; Alharthy, R.D.; Zubair, M.; Ahmed, M.; Hameed, A.; Rafique, S. Toxic and heavy metals contamination assessment in soil and water to evaluate human health risk. *Sci. Rep.* **2021**, *11*, 17006. [CrossRef]
72. Pan, L.; Wang, Y.; Ma, J.; Hu, Y.; Su, B.; Fang, G.; Wang, L.; Xiang, B. A review of heavy metal pollution levels and health risk assessment of urban soils in Chinese cities. *Environ. Sci. Pollut. Res. Int.* **2018**, *25*, 1055–1069. [CrossRef] [PubMed]
73. Pena-Fernandez, A.; Gonzalez-Munoz, M.J.; Lobo-Bedmar, M.C. Establishing the importance of human health risk assessment for metals and metalloids in urban environments. *Environ. Int.* **2014**, *72*, 176–185. [CrossRef]
74. Lu, C.; Qing, L.; Wang, X.; Zhang, C.; Xi, Z.; Liu, Z.; Wang, X. Characterization and Risk Assessment of Heavy Metals in Soil of Mine Area in the Yunnan-Guizhou Area. *Environ. Sci.* **2024**, 1–17. (In Chinese) [CrossRef]
75. Yang, Q.; Li, Z.; Lu, X.; Duan, Q.; Huang, L.; Bi, J. A review of soil heavy metal pollution from industrial and agricultural regions in China: Pollution and risk assessment. *Sci. Total Environ.* **2018**, *642*, 690–700. [CrossRef]
76. Guo, S.; Zhang, Y.; Xiao, J.; Zhang, Q.; Ling, J.; Chang, B.; Zhao, G. Assessment of heavy metal content, distribution, and sources in Nansi Lake sediments, China. *Environ. Sci. Pollut. Res. Int.* **2021**, *28*, 30929–30942. [CrossRef]
77. Ihedioha, J.N.; Ogili, E.O.; Ekere, N.R.; Ezeofor, C.C. Risk assessment of heavy metal contamination of paddy soil and rice (Oryza sativa) from Abakaliki, Nigeria. *Environ. Monit. Assess.* **2019**, *191*, 350. [CrossRef]
78. Yang, S.; Zhao, J.; Chang, S.X.; Collins, C.; Xu, J.; Liu, X. Status assessment and probabilistic health risk modeling of metals accumulation in agriculture soils across China: A synthesis. *Environ. Int.* **2019**, *128*, 165–174. [CrossRef] [PubMed]
79. Jin, J.; Zhao, X.; Zhang, L.; Hu, Y.; Zhao, J.; Tian, J.; Ren, J.; Lin, K.; Cui, C. Heavy metals in daily meals and food ingredients in the Yangtze River Delta and their probabilistic health risk assessment. *Sci. Total Environ.* **2023**, *854*, 158713. [CrossRef] [PubMed]
80. Gu, Y.; Lin, Q.; Gao, Y. Metals in exposed-lawn soils from 18 urban parks and its human health implications in southern China's largest city, Guangzhou. *J. Clean. Prod.* **2016**, *115*, 122–129. [CrossRef]
81. Wu, J.; Wang, H. Assessment and amendment methods of heavy metal non-carcinogenic health risks in agricultural land around smelters. *J. Environ. Eng. Technol.* **2024**, *14*, 112–120. (In Chinese) [CrossRef]

MDPI AG
Grosspeteranlage 5
4052 Basel
Switzerland
Tel.: +41 61 683 77 34

Toxics Editorial Office
E-mail: toxics@mdpi.com
www.mdpi.com/journal/toxics

www.ingramcontent.com/pod-product-compliance
Lightning Source LLC
Chambersburg PA
CBHW061630210926
51850CB00009B/2504
* 9 7 8 3 7 2 5 8 5 7 6 3 0 *